THE MONTY HALL PROBLEM

& OTHER PUZZLES

IVAN MOSCOVICH

DOVER PUBLICATIONS, INC.
MINEOLA, NEW YORK

To Anitta, Hila, and Emilia, with love

Bibliographical Note

This Dover edition, first published in 2011, is an unabridged republication of the work originally published in 2004 by Sterling Publishing Company, Inc., New York.

Library of Congress Cataloging-in-Publication Data

Moscovich, Ivan.
    The Monty Hall problem and other puzzles / Ivan Moscovich. — Dover ed.
        p. cm.
    Originally published: New York : Sterling Publishing Co., 2004
    ISBN-13: 978-0-486-48240-8
    ISBN-10: 0-486-48240-5
    1. Puzzles I. Title
GVI493.M6135 2011
793.73--dc22

                                                            2011002855

Manufactured in the United States by Courier Corporation
48240501
www.doverpublications.com

Contents

Introduction

Ever since my high-school days I have loved puzzles and mathematical recreational problems. This love developed into a hobby when, by chance, some time in 1956, I encountered the first issue of *Scientific American* with Martin Gardner's mathematical games column. And for the past 50 years or so I have been designing and inventing teaching aids, puzzles, games, toys and hands-on science museum exhibits.

Recreational mathematics is mathematics with the emphasis on fun, but, of course, this definition is far too general. The popular fun and pedagogic aspects of recreational mathematics overlap considerably, and there is no clear boundary between recreational and "serious" mathematics. You don't have to be a mathematician to enjoy mathematics. It is just another language, the language of creative thinking and problem-solving, which will enrich your life, like it did and still does mine.

Many people seem convinced that it is possible to get along quite nicely without any mathematical knowledge. This is not so: Mathematics is the basis of all knowledge and the bearer of all high culture. It is never too late to start enjoying and learning the basics of math, which will furnish our all-too sluggish brains with solid mental exercise and provide us with a variety of pleasures to which we may be entirely unaccustomed.

In collecting and creating puzzles, I favour those that are more than just fun, preferring instead puzzles that offer opportunities for intellectual satisfaction and learning experiences, as well as provoking curiosity and creative thinking. To stress these criteria, I call my puzzles Thinkthings.

The *Mastermind Collection* series systematically covers a wide range of mathematical ideas, through a great variety of puzzles, games, problems, and much more, from the best classical puzzles taken from the history of mathematics to many entirely original ideas.

This book, *The Monty Hall Problem & Other Puzzles*, contains the famous game show problem, based on seemingly very simple principles of probability. This provoked an enormous uproar among mathematicians. Understanding the theory behind this puzzle will help you solve many of the others in the book.

A great effort has been made to make all the puzzles understandable to everybody, though some of the solutions may be hard work. For this reason, the ideas are presented in a novel and highly esthetic visual form, making it easier to perceive the underlying mathematics.

More than ever before, I hope that these books will convey my enthusiasm for and fascination with mathematics and share these with the reader. They combine fun and entertainment with intellectual challenges, through which a great number of ideas, basic concepts common to art, science, and everyday life, can be enjoyed and understood.

Some of the games included are designed so that they can easily be made and played. The structure of many is such that they will excite the mind, suggest new ideas and insights, and pave the way for new modes of thought and creative expression.

Despite the diversity of topics, there is an underlying continuity in the topics included. Each individual Thinkthing can stand alone (even if it is, in fact, related to many others), so you can dip in at will without the frustration of cross-referencing.

I hope you will enjoy the *Mastermind Collection* series and Thinkthings as much as I have enjoyed creating them for you.

—Ivan Moscovich

In this book we'll be testing all kinds of skills—observational, logical, practical, and unusual! This initial pair of teasers gives you a taste of things to come by demonstrating that sometimes a lateral approach may get you further than pure number-crunching.

▼ INEBRIATED INSECT

Two ladybugs landed on my glass: one outside, exactly in the middle of the glass, the other exactly opposite but inside the glass. The height of the glass is $5\frac{1}{2}$ units and the width is 4 units. Can you describe the shortest path from one to the other? Are you able to estimate how long that path is?

ANSWER: PAGE 98

▼ THE SHORT AND LONG OF IT

The schematic diagram below shows the road network between a set of towns.

Mathematically, it is called a tree graph. The colors of the roads indicate how long they are to the nearest 5 miles.

(a) Can you find the longest possible distance between any two towns, without taking any U-turns?

(b) Generally, this kind of problem can challenge even the most advanced computers, especially if the number of towns increases. Can you describe a simple way to demonstrate which path through the network is longest without resorting to any math?

ANSWER: PAGE 98

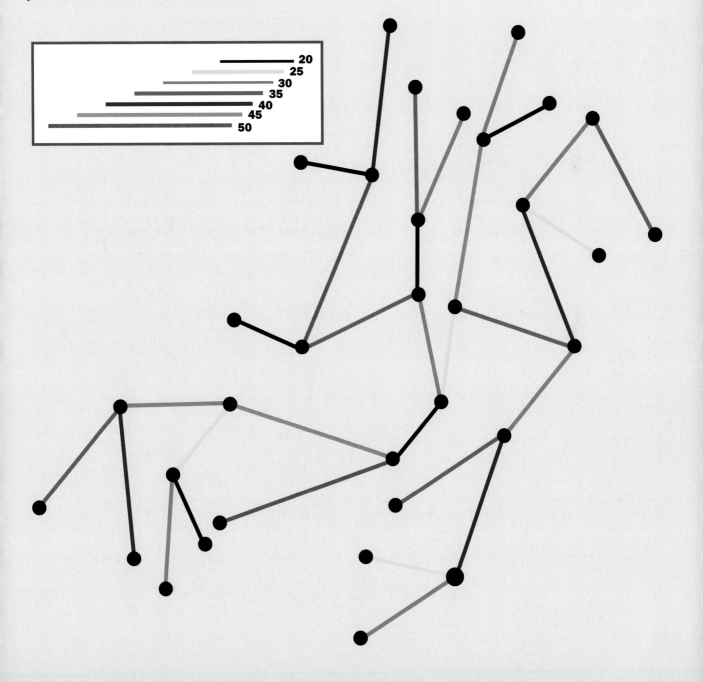

The most important thing a person inherits is the ability to learn a language. Language—especially written language—makes connection possible between people living in vastly different circumstances, places, and times. What human beings know of the past and can foretell of the future comes from language.

? DID YOU KNOW?

Scientists can now write using a method which is so small that everything that's been written by man since the beginning of time could be written on a piece of paper about the size of your living room carpet.

✳ Communication and mathematics

To get a true sense of the significance of language, consider this: Is it possible to obtain meaning from something without using words or signs? Some philosophers believe that a world without language would be a world devoid of meaning.

Language is carried visually either by signs, written marks that stand for units of language, or symbols, which represent an object itself.

In the 20,000 years since human beings first scratched simple tallies on bones, the visual aspect of language has flourished. First objects, then words, were abstractly represented. By 300 B.C. the library of Alexandria contained some 750,000 papyrus scroll books, the greatest storehouse of knowledge the world has ever seen—this was made possible only through the use of signs and symbols.

Later, technological developments such as block printing (by the Chinese) and movable type (by Johannes Gutenberg) enabled written language to reach virtually every person on the planet. Symbolic language promotes a type of visual thinking that today's designers and communication engineers must take into account.

Older ways of presenting complex ideas and more verbal forms of recalling information are quickly rendered obsolete. Change happens so quickly that even written language may not be the most trustworthy means of communicating with future generations. It is no exaggeration to say that anyone trying to send a message to the future—be it a memorial to a great leader or a warning about a toxic waste site—ought to look at the efforts that have been made by astronomers to communicate with intelligent life forms on other planets.

If such aliens existed, they would be unfamilar with any human language, written or spoken. Astronomers involved with the Search for Extra-Terrestrial Intelligence (SETI) are scanning the heavens with radio telescopes in search of a scrap of message—intentional or accidental—amid the natural noise of the stars, although no one knows what such a message might look like.

Other astronomers have tried to send messages to distant stars in the form of pictographs symbolizing everything from the human form to the lightest chemical elements. But even such seemingly simple pictures would require some ingenuity to decode them. Perhaps mathematics will provide the key. Only mathematics can be a language universal enough for both human beings and extra-terrestrials to understand, as the puzzles on page 9 demonstrate. The interstellar greeting may not therefore be "hello," as you might think, but "one, two, three ..."

▶ GET THE MESSAGE?

Messages like these were sent to outer space in order to establish communication with intelligent life, beings who would be unlikely to understand our written or spoken languages.

For this reason, interstellar messages can use only mathematical language and binary codes to send messages into outer space with a chance of establishing contact. Can you decipher what the message was?

ANSWER: PAGE 99

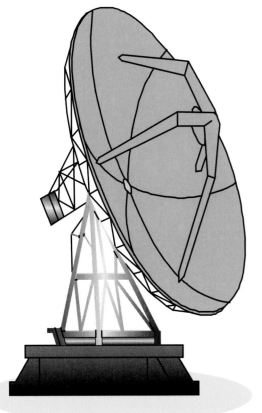

▲ THE FIRST CONTACT

Mathematics is the only universal language that we could expect intelligent aliens to understand.

The best way to establish communication with aliens is by mathematics: just counting … like in our message above. But the question here is: Did they understand the message?

ANSWER: PAGE 99

Early human beings learned how to build structures more efficiently by simple trial and error. When the ancients added a great deal of ingenuity, they accomplished wondrous feats of architecture and engineering—in the process developing the first form of geometry.

PREHISTORIC CAVE PAINTING
The number of hunters in the team doesn't match the number of animals killed. This fact must have led to man's first mathematical ideas and problems—the concepts of MORE and LESS.

✳ Prehistoric and early geometry

Mathematical forms are present in plant life. Certain leaves, millions of years ago, arranged themselves about a stalk in obedience to the law of series first expressed by Leonardo Fibonacci. We know today the law involved, but how did the plants come to know it?

Was it the result of countless experiments in search of maximum efficiency, culminating in a plant habit, or was some other law involved that only the future will reveal?

With the appearance of animal life came the possibility of the recognition of mathematical concepts such as form, number, and measure. Evidence thtat spiders, for example, recognize regular polygons, logarithmic spirals, and similarity of figures can be seen in any spider's web. The hexagonal structure of a bee's honeycomb provides the best volume available for the least amount of wax walls constructed.

With the advent of the human race there developed an opportunity for mathematics to show itself more consciously. Art contributed to an appreciation of geometry, religious mysticism and commerce contributed to the development of numbers, and each influenced the creation of an interest in architecture and astronomy.

Dramatic natural events, such as solar eclipses, lightning, and the paths of the sun and moon, aroused man's inborn curiosity. Much later, when that curiosity was disciplined by measurement and reinforced by mathematical reasoning, we became scientists.

Even the cavemen who traded spears for animals were using arithmetic. They probably counted out on the fingers of their hands and maybe the toes of their feet. Later, people began to own many things and had to keep count of them by making notches on sticks or trees (tally

marks). Tally sticks were also used as calendars to mark the passage of hours and days.

From here, numbers were invented. At first, they looked very much like tally marks. But with all the ancient number systems, it was very difficult to perform even the most simple arithmetical operations. A mechanical method of computation was necessary. The Romans performed arithmetic by counting pebbles, which was the most primitive form of the abacus.

Geometry as a science, however, was born in ancient Greece. Greek geometers were absorbed in the study of simple forms—the circle, the square, the triangle. Armed with only compasses and straightedges, they set out to find geometric truths; by 350 B.C., Euclid had compiled a set of rules concerning space and shapes that dominated geometry for 2,000 years.

❋ Geometry and the art of puzzling

How did puzzles develop? They have a surprisingly long history. Pythagoras, credited with the right-angled triangle theorem, provided the basis of endless geometrical puzzles (the Chinese already knew about the Pythagorean theorem in the 1st century A.D.). And Archimedes, the greatest mathematician of antiquity, created and solved recreational problems, of which the most famous is the "Eureka" incident, associated with the golden crown of Hieron II, the king of Syracuse.

In the late 8th century, Alcuin, an English scholar, introduced the river-crossing problem, which today has an endless number of variations. His main work, *Problems to Sharpen the Young*, is the earliest known collection of problems in Latin.

In A.D. 1175, Leonardo of Pisa (Fibonacci) discovered the Fibonacci number sequence, which proved to be of enormous importance in many fields of science and mathematics.

The 17th century produced books devoted solely to recreational problems not only in mathematics but in physics as well. The first important contribution was that of the Frenchman Claude-Gaspar Bachet de Méziriac, who is remembered chiefly for two works: his *Diophanti*, the first edition of a Greek text on the theory of numbers (1621), and his *Problèmes Plaisants et Délectables qui se Font par les Nombres* (1612). The latter title was the forerunner of similar collections to follow. The emphasis here was more on arithmetic rather than on geometric puzzles.

In 1624 a French Jesuit, Jean Leurechon, writing under the pen name of van Etten, published his *Récréations Mathématiques*, a book that passed through at least 30 editions.

And in Germany, Daniel Schwenter, a professor of Hebrew and Oriental languages and mathematics, compiled the immensely popular *Mathematische und Philosophische Erquickstunden* in 1636.

Charles Bombaugh, in an 1874 book, *Gleanings for the Curious*, included a section on puzzles. He wrote "such puzzles are not work but play. Solving them gives quickness of thought and facility in turning about a problem every way, viewing it in every possible light."

But popular interest in recreational math and science only really became strong about a century ago thanks to Sam Loyd, America's foremost puzzle inventor, and Henry Ernest Dudeney, his British counterpart. And in modern times it has acquired an enormous boost through the work of Martin Gardner, and lately by his followers, the mathematician Ian Stewart and others.

PLATO (427–347 BC)

In about 387 B.C., the Greek scholar Plato founded a school of learning in Athens which he called the Academy—an institution devoted to philosophy, the sciences, and mathematics.

In his Theory of Forms, *Plato considers mathematical objects as perfect forms. In his* Phaedo *and* Republic, *he talks of objects in the real world in their mathematical abstract forms. Although Plato made no mathematical discoveries himself, his school and belief that mathematics provides the finest training for the mind was of crucial importance for the further development of the subject.*

ΜΗΔΕΙΣ ΑΓΕΩΜΕΤΡΗΤΟΣ
ΕΙΣΙΤΩ ΜΟΥ ΤΗΝ ΣΤΕΓΗΝ

"Let no one ignorant of geometry enter.**"**
Tablet over the door of Plato's school

❓ DID YOU KNOW?

The Latin word for pebble is "calculus."

The usual geometry system we learn at school (with the x, y, and z axis) is called Euclidean geometry. But what if we use a different system? One approachable non-Euclidean geometry is the so-called "taxicab geometry," which you can explore with paper. It was first seriously researched by Hermann Minkowski, a Russian mathematician, who taught the young Einstein in Zurich.

✳ Non-Euclidean geometries

Geometry has its roots in the measuring of parcels of land. Indeed, the literal translation of geometry is "measurement of the earth." But the complexity of the shapes of fields and the ingenuity of ancient geometers employed to perform their jobs soon led geometry away from land surveys and toward the study of the relationships of abstract shapes.

For some 2,000 years or so, Euclid's geometry was generally considered to be the only possible geometry, and this despite the fact that no one actually knew why his theorems were true. But since Euclid's theorems concerning straight lines and two-dimensional planes were self-evident—in other words, they could actually be seen to work—no one thought to challenge them.

Early in the 19th century, however, mathematicians discovered that not only were many of Euclid's tenets not self-evident but they weren't always true. For instance, when they substituted a sphere for Euclid's plane, mathematicians found that, contrary to Euclid's axiom, two parallel lines *would* meet, just like two lines of longitude join at the Poles.

Why did it take so long to overturn Euclidean geometry? The answer is that individual human beings occupy a tiny area of a huge sphere (the Earth) and in such a limited area Euclidean geometry works well. For nearly all practical purposes, ar-

chitects can design and build structures as though the Earth were flat. Indeed, in the ancient world there were few practical applications for any geometry that wasn't Euclidean. But non-Euclidean geometries have powerful and important applications in modern physics and cosmology.

Astronomers, for example, make use of non-Euclidean geometry to describe the paths of light rays as they bend around very heavy bodies, such as stars, black holes, and galaxies. Combined with Einstein's theories of relativity, non-Euclidean geometry has solved many mysteries about space and time that could not be tackled in any other way.

TAXICAB GEOMETRY
The geometry of Gridlock City

Imagine Gridlock City, in which the streets run either north-south or east-west. (Many cities established in the 19th century possess just that sort of grid.) To get around Gridlock City by taxicab, one must measure distances not "as the crow flies" but "as the cab drives"—that is, along the lines of the square grid.

Taxicab distances are in general longer than ordinary distances except when you drive from one end of the street to the other. If Gridlock City is made up of straight lines on a plane, how can it be non-Euclidean?

One of Euclid's axioms states that the shortest distance between two points is a straight line. Is that the case in Gridlock City? In fact, the shortest path in most cases is a series of short lines, since travel is restricted to the street grid. You must drive around the blocks, not through them. Does that mean it is impossible to have a circle in Gridlock City? By definition a circle is a shape in which all points are equidistant from a fixed point.

Suppose that there are six blocks to a mile in Gridlock City and you travel a mile by taxi from the center of the city. Where do you end up?

You could travel six blocks due east and stop. Or you could go five blocks east and one block north, or four blocks east and two blocks north. All those points lie on the "taxicab circle" of radius 1 mile.

Can you plot the shape of such a circle (see page 16)? Since the "straight lines" (the shortest paths) of taxicab geometry may be crooked from the Euclidean point of view, the concept of an "angle" becomes meaningless or different in this geometry, and therein lies the challenge.

It is nonetheless possible to define close analogs of Euclidean polygons, including a two-sided polygon called a "biangle," which would be impossible in Euclid's geometry.

▶ BIANGLES

A two-sided polygon called a "biangle," non-existent in Euclidean geometry, can exist in taxicab geometry. A biangle of length eight units between two points is shown. It should be obvious that different biangles can share the same pair of "corner" points, and the two "sides" of any biangles must be equal because they join the same two points. How many biangles of the same size as the one shown can you find?

ANSWER: PAGE 99

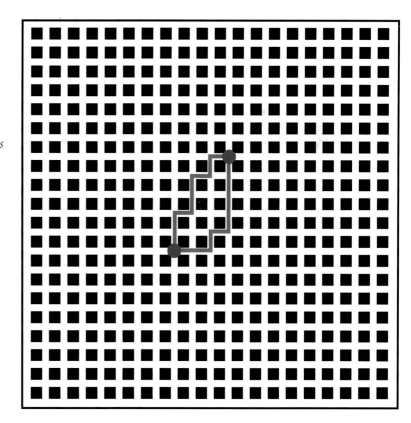

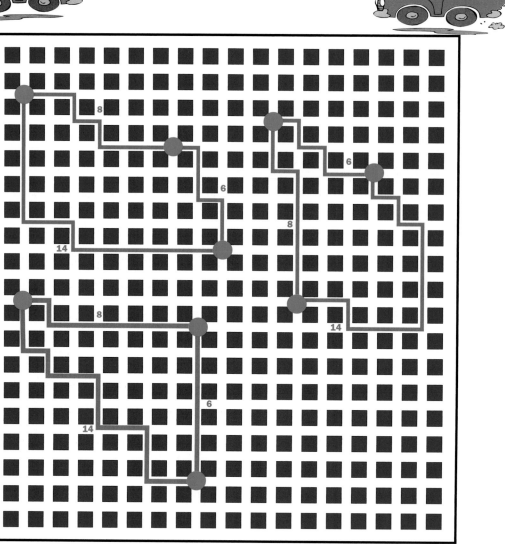

▲ TAXI TRIANGULATION

In Euclidean geometry, given three points there can be only one triangle. In taxicab geometry, however, there can be more than just one triangle with sides of, say, 6, 8, and 14 units, as shown. Among the many triangles with such sides, can you find the one that has the smallest area (that is, the one that encompasses the smallest number of block units)?

ANSWER: PAGE **99**

▼ SQUARE ROUTE

*As with triangles, there can be many squares in taxicab geometry; for
example, three squares of side 6 are shown. Can you work out the area
of the smallest square for each of the three squares of sides 6 block units?
You may change the lines but not the four points in each given square.*

ANSWER: PAGE *100*

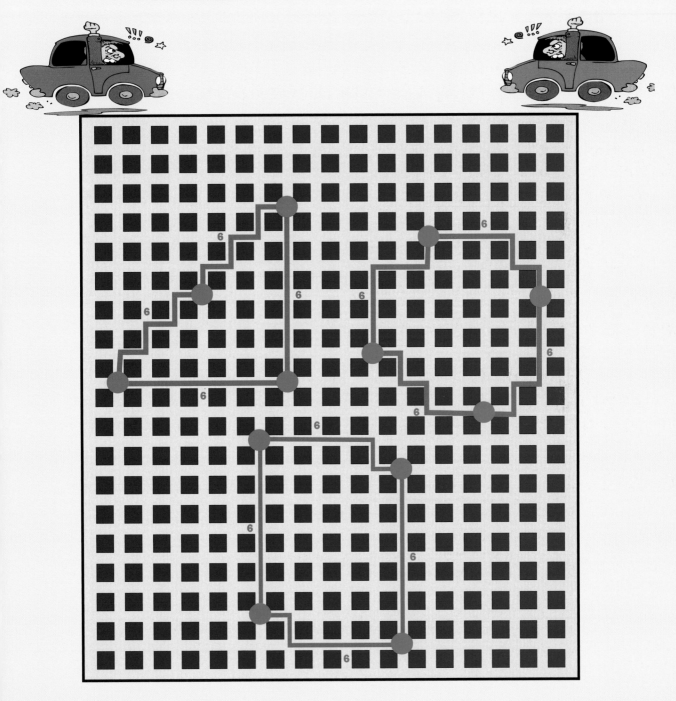

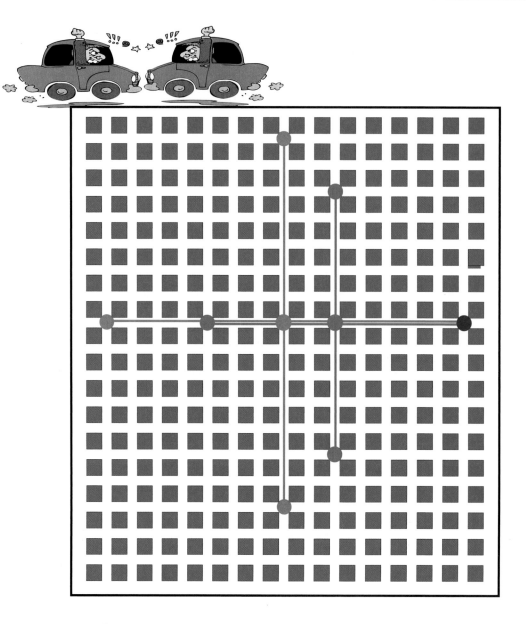

▲ GOING ROUND IN CIRCLES

In Gridlock City you can move around only along blocks from one point to another. By definition a circle is a shape in which all points are equidistant from a fixed point—the center. Can you plot two circles: one with its center at the red point and with a radius of 7 block units; and the other, from the green point, with a radius of 5 block units? At how many points will the two circles intersect? The two main diameters of the two circles are shown (with the first point of intersection marked in blue).

ANSWER: PAGE 100

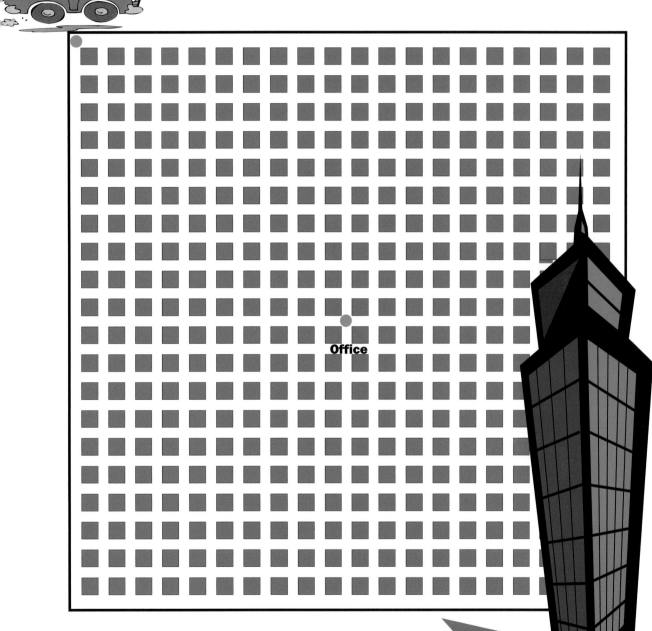

▲ ALL ROADS LEAD TO...

You live at the top left corner of the map and work in an office at the point shown. What is the shortest path to your office and how many different routes of the shortest length can you take?

Hint: Start with a smaller journey then look for a rule.

ANSWER: *PAGE 101*

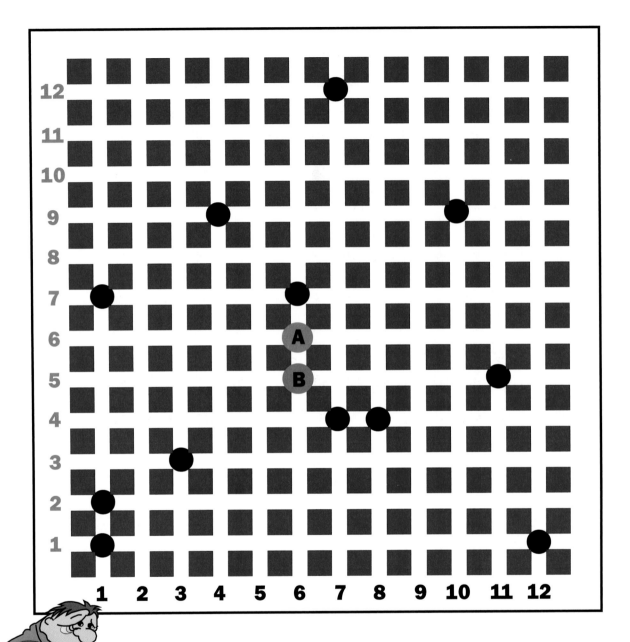

▲ LOCAL LIQUOR

A bar frequently visited by 12 friends living in the city of Gridlock is situated at a spot minimizing the total walking distance for all of them taken together. At which of the two marked spots is the bar situated? A or B?

ANSWER: PAGE 102

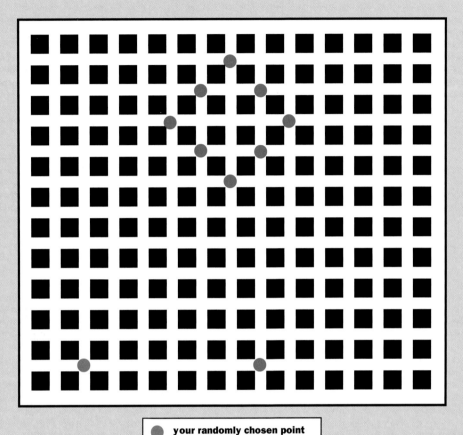

● your randomly chosen point

● the center of the circle on
which the treasure is hidden

● given circle

▲ TREASURE HUNT

Treasure is hidden at one of the intersections of the city of Gridlock. You have randomly chosen an intersection and have received secret information: The treasure is hidden on a taxicab geometry circle of 7 blocks radius, the center of which is 6 blocks east from your chosen point on one of the intersection points with the given green circle. How many trial digs are needed to pinpoint the intersection where the treasure is hidden?

ANSWER: PAGE 102

```
B J O U O P O X E L A X C D A S T U W
Q R W H I A L A L R L B A E Z F G H O
Y E D S T C E L A O J L Y K L O O P E
B C E A O A A M O A N I L I Q W E O E
D E D E K K O U A B M E K L L Y E A Z
C Z G O O W A I ! N E J O X X D R U E
F A F F A T J W F Z I T Q P S S M B I
A G G O H R O C X Q A D R H G C C F I
O O O N B N F T D Y B K F A E U H H I
M S G J - E R Y I X C A C E Y W X S S
L L C K V A K U A O I C L E A E T T Y
Z X C F E V K A E P G J K D I J D R Z
A Q Y O C C E Q G X A E A C E - O P Q
H W T E L A M D U E I S D W L G B M N
A K G X C D O C G A E P G E Q T Y K L
J W R S H E I C C I U G A C E E A T I
U A B A E C E R A Z D V A U V U N M E
Y T A W N I P B G U E F G H G I H I O
K O U Z C J L E S - D S U K R I B V I
```

▲ CUBIC CRYPTOGRAM

A message has been left for the Gridlock spy. It says that the hidden message can be read along the border of a gridlock circle of radius 7. However, the instructions did not indicate where the center of the circle is, just that it lies somewhere in the middle column of letters.

Of course, the spy is familiar with the geometry of Gridlock City, in which distances can be measured only around blocks, from junction to junction. What did the message say?

ANSWER: PAGE **103**

▼ DELIVER THE GOODS

In Gridlock City the pizza boy must make five deliveries from the pizza shop located at point 1 to five other locations (2, 3, 4, 5, and 6 as shown). The blocks in Gridlock City are all of the same size. What will be his best route starting from the pizza shop and returning again, visiting each location just once but in any order?

Here's a suggested approach for you to try: Work out the minimum distances between each set of locations, then find out which route will allow you to visit them all in the smallest taxicab distance.

ANSWER: PAGE **103**

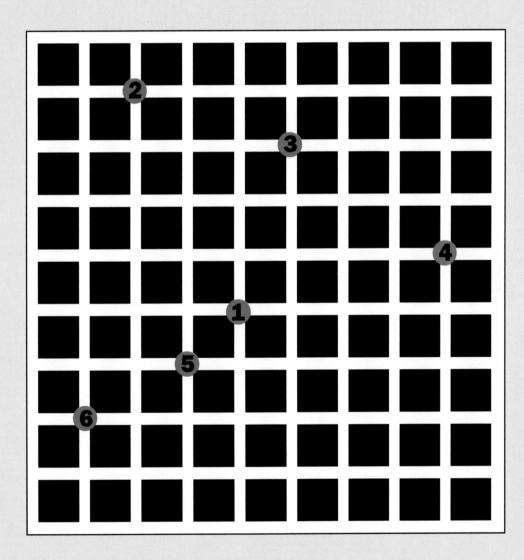

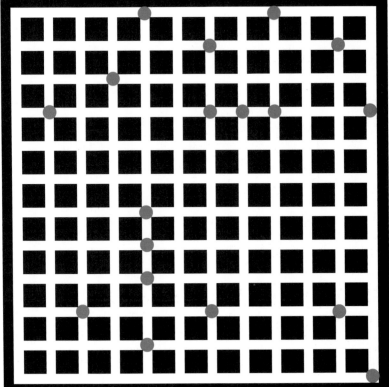

▶ SCENIC TOUR

Can you create a closed journey through Gridlock City that passes through every intersection, except those that are blocked by the red traffic signs? You must finish at the same junction from which you started.

ANSWER: PAGE **104**

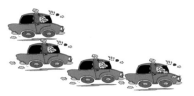

▶ RIGHT THIS WAY

Getting across town by car can be a nightmare, not because of the traffic but because the crazy road signs always seem to force you to go where you don't want to.

In the town of Gridlock, the problem is even worse: The town traffic authorities have increased the number of signposts, and have even invented some new ones, so that at most intersections there is at least one way you cannot turn. Getting from one side of town to the other therefore now involves some surprising twists and turns. Can you find routes across town—beginning on the left and ending on the right—following the road signs at the exit of every street at each junction, and connecting the red entry with the red exit; the blue entry with the blue exit; and the green entry with the green exit?

Note: If given more than one option, you may choose which direction to take.

ANSWER: PAGE **104**

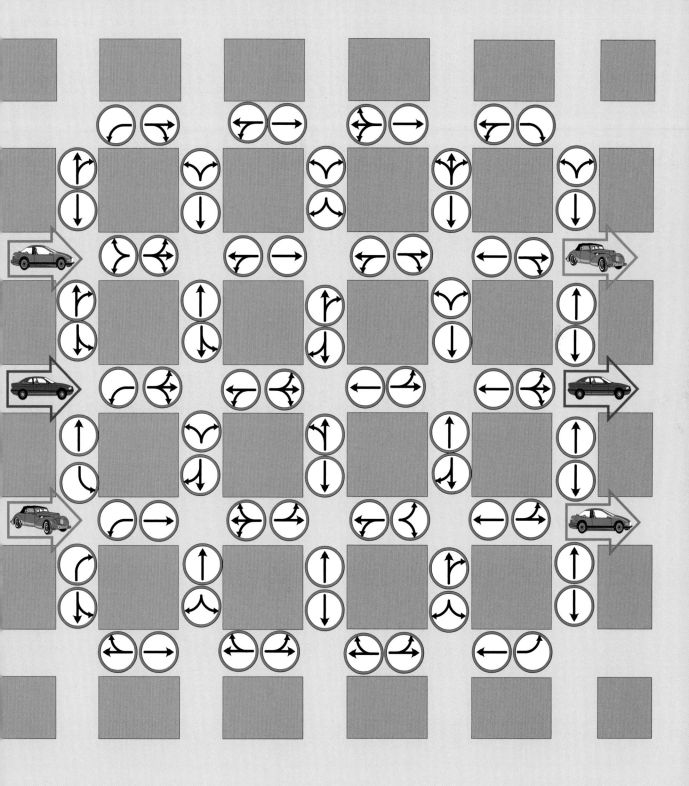

In the past, some people have had trouble coming to terms with the fact that pi (π), the ratio between a circle's diameter and circumference, is what it is (approximately 3.14). In 1897, Indiana's State Legislature tried to pass a law that set it at 3.2. Fortunately for them, the constant of pi is above the law…

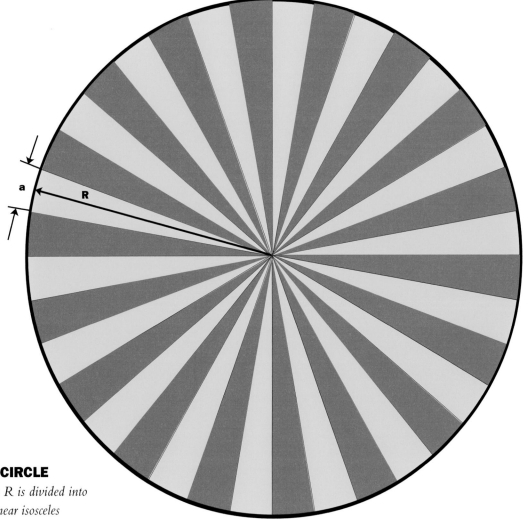

▶ AREA OF A CIRCLE

The circle of radius R is divided into a great number of near isosceles triangles with sides R and bases "a" (small circular arcs approximating straight lines).

Can you work out the area of the circle like Archimedes did a long time ago?

ANSWER: PAGE *105*

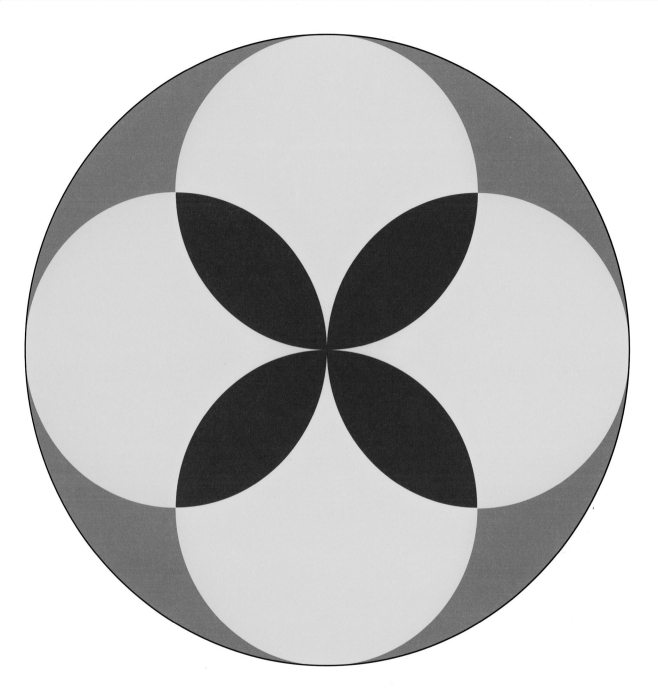

▲ CLOVER

Look at this image of a clover leaf. Can you work out which area is greater, the red or blue?

ANSWER: PAGE 105

No point in the universe is truly fixed. A point that remains stationary within a car may be tracing a linear path as the car speeds down the highway. A point on a mountain follows the earth around the sun. And even the sun and the Milky Way galaxy have their own paths through an ever-expanding universe. The motion of a fixed point on a moving body traces a curve that can have very unusual properties. For example, the curve traced by a point on a rotating circle is called a cycloid.

The cycloidal curve appears in many places in modern society: Mechanical gears have teeth whose sides possess a cycloidal curve; a machine engraves an elaborate cycloid on the plates used for printing banknotes; a popular science toy known as the Spirograph produces an endless variety of cycloidal shapes with just a very few moving parts. Other similar curves include the spiral and the involute—the line traced by the end of a taut thread as it is unwound off a spool.

▶ **4-WAY RACE**

Four identical balls are released at the same time on four different tracks: straight, bent, circular, and cycloidal, as shown (in black lines). Can you tell which ball will be the first to arrive at the end of the slope?

One of these shapes is called the "brachistochrone," the special curve on which an object descending under gravity will be faster than on any other curve. (This idea is the basis of Einstein's account of space as curved.)

ANSWER: PAGE *106*

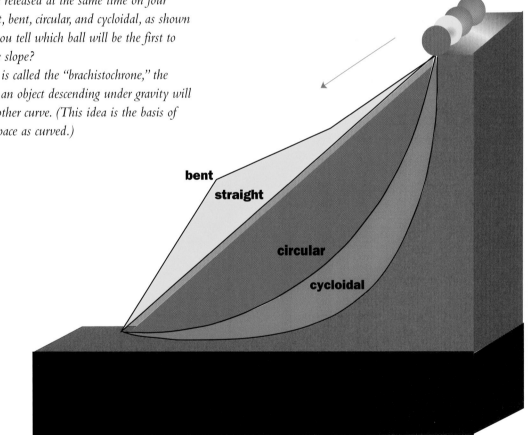

bent

straight

circular

cycloidal

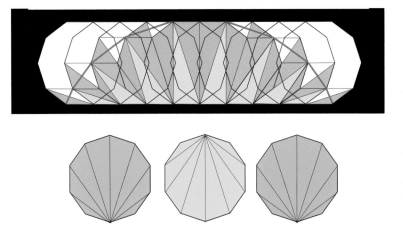

◄ POLYGONAL PROOF

Polygonal cycloids are good analogies for obtaining solutions about the length and area of a cycloid. Can you work out the area under the polygonal arch (in red) generated by the revolving regular decagon along a straight line?

This beautiful visual proof was created by Philip R. Mallinson (Roger B. Nelsen, Proofs Without Words II, *The Mathematical Association of America, 2000).*

ANSWER: PAGE **106**

Cycloid and cycling

In 1882, the German mathematician Ferdinand von Lindemann proved that it is not possible to construct a square equal in area to a given circle if only a compass and straightedge are available.

On the other hand, it is possible to square the circle if a certain mechanism, consisting of a circle rolling along a straight line, is used. This produces a curve called a cycloid. As the circle completes one revolution, the point A on its circumference, moving from A to Z, describes a cycloid. The length of the straight line from A to Z is equal to the circumference of the circle—that is, $2\pi r$.

Thus, if B is the midpoint of AZ, then $BZ = \pi r$. Hence, if $CZ = r$, the area of the rectangle BZCD is $\pi r \times r = \pi r^2$, which is also the area of the rolling circle. It is now possible to cut the rectangle in half and rearrange it to form a square of side ZE whose area is equal to the circle.

A   B   r Z   E

D   C

The equiangular, logarithmic, or Bernoulli Spiral has a longer history than the recorded science of mathematics. It appeared millions of years ago in spiral galaxies, nautilus shells, in the arrangement of sunflower seeds, and many other natural phenomena. In mathematics, the Frenchman René Descartes was the first to discover its formula around the middle of the 17th century. It was later studied by Jacques Bernoulli, Evangelista Torricelli, and others. Equiangular spirals are formed by pursuit curves. The equiangular spiral has a special property such that if you were to draw a straight line from the center outward, that line would always meet the spiral at the same angle.

▶ TRIANGULATION: The puzzle

Start with a small right-angled triangle of 1-unit sides, add a further right-angled triangle to its hypotenuse, and so on for further triangles in succession. How will the resulting curve look, and what will the size of the hypotenuses of the added right-angled triangles be? The fascinating answer is on page 29 opposite.

" *In the growth of a shell we can conceive no simpler law than this: Namely that it shall widen and lengthen in the same unvarying proportions: and this simplest of laws is what Nature tends to follow. The shell, like the creature within it, grows in size but does not change its shape; and the existence of its constant relativity of growth, or constant similarity of form, is of the essence, and may be made the basis of a definition, of the equiangular spiral.* **"**
D'Arcy Thompson, 1961

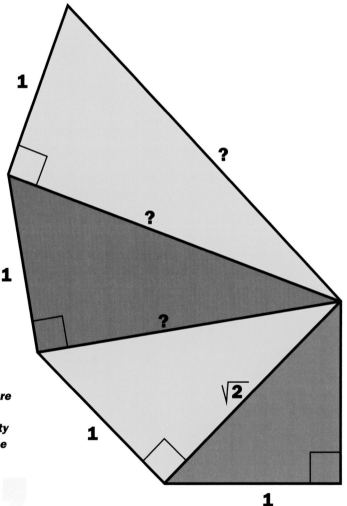

▼ TRIANGULATION: The answer

If we use the Pythagorean theorem on each triangle (on page 28) in turn, we obtain the following pattern:

$$\sqrt{(1^2 + 1^2)} = \sqrt{2}$$
$$\sqrt{(1^2 + (\sqrt{2})^2)} = \sqrt{3}$$
$$\sqrt{(1^2 + (\sqrt{3})^2)} = \sqrt{4}$$
$$\sqrt{(1^2 + (\sqrt{4})^2)} = \sqrt{5} \ etc.$$

So we can see how the lengths of the hypotenuses of each triangle grow each time. This is why this shape is called the Square Root Spiral. If we continue the shape many times, we obtain the beautiful shape below.

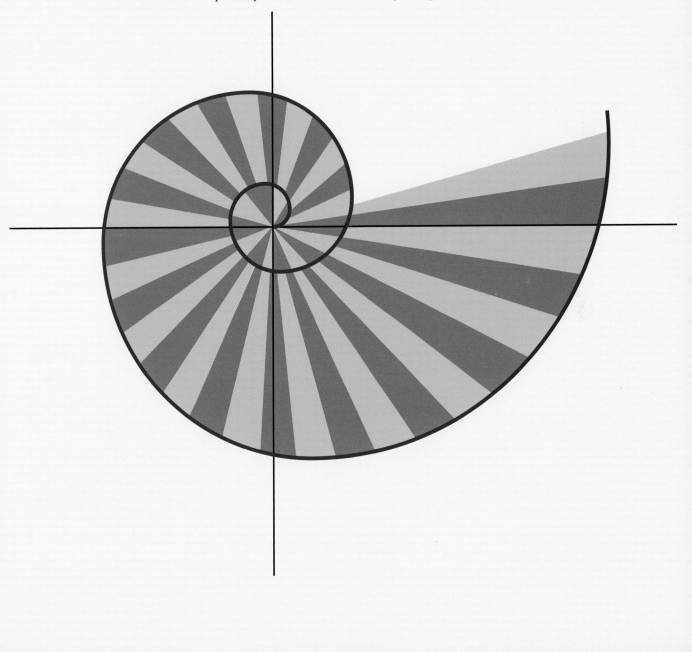

In this set of geometrical problems, you'll need to think about how to define the center of a triangle. In fact, there are several possible ways. If you guess right, you'll receive a police commendation in no time at all.

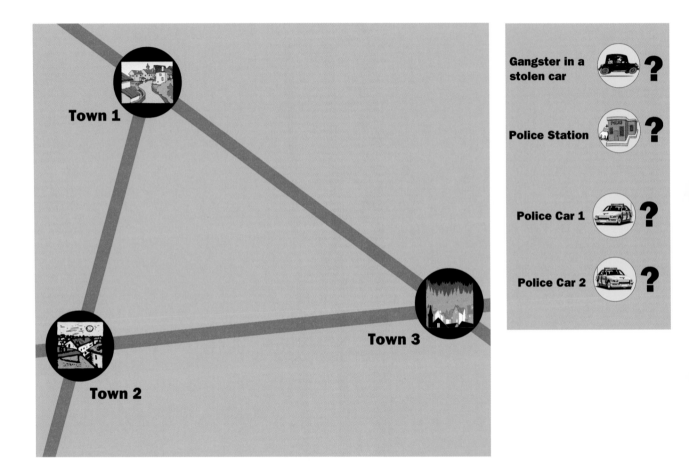

▲ POLICE CHASE

Puzzle 1 The three towns shown above, interconnected by three main roads, will be patrolled by officers from a police station to be located at a strategic point inside the triangle, so that it is at a minimal total distance from the three towns.

Can you locate the point where the police station should be built?

Puzzle 2 A gangster in a stolen car is resting overnight in a motel in one of the three towns.

Where should Police Car 1 position itself in the middle of the triangle so that, if the gangster blows his cover, the policeman can drive the same distance to any town?

Puzzle 3 Where should Police Car 2 position itself inside the triangle so that, if the gangster evades capture, it is the same distance away from any of the three roads?

ANSWER: PAGE 107

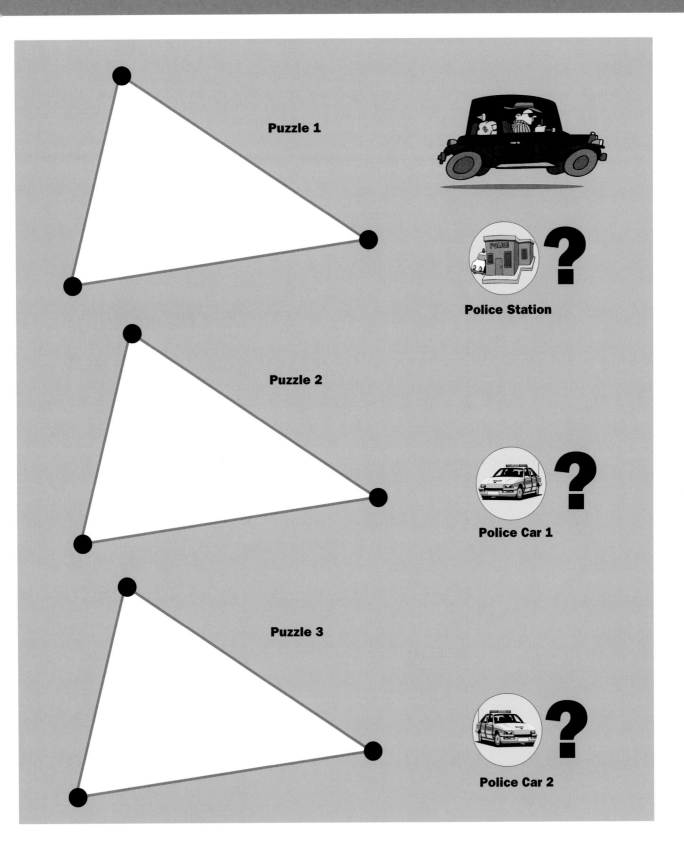

Puzzle 1

Puzzle 2

Puzzle 3

Police Station

Police Car 1

Police Car 2

Murano is a small island in Venice's Laguna Veneta, which is internationally famous for its glass-blowing. So enjoy looking at these geometric vases, but remember—you'll pay for any damages!

▼ BRIGHT BLOSSOMS

The four Murano glass vases on page 32 are blown from an identical block of glass as displayed, although the inside spaces are different, as shown. Can you estimate what the water levels of the four vases will be when the water in each is poured into the lower container?

ANSWER: PAGE **108**

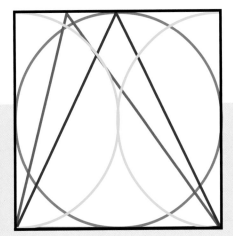

100%

— 75%

—50%

— 25%

—0%

Did you know that a flea is capable of jumping over one foot (30cm) in a single bound? That's about the same as a human being leaping over 200 yards (over 180 meters). So don't delay before trying these puzzles—jump to it!

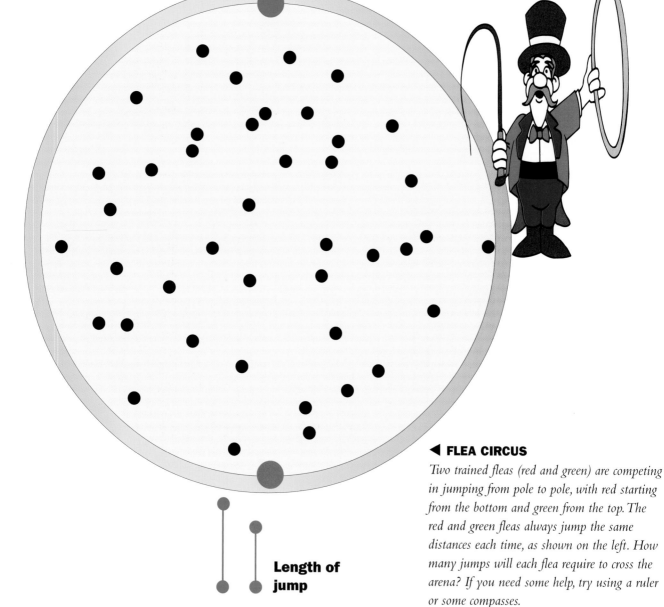

Length of jump

◄ FLEA CIRCUS

Two trained fleas (red and green) are competing in jumping from pole to pole, with red starting from the bottom and green from the top. The red and green fleas always jump the same distances each time, as shown on the left. How many jumps will each flea require to cross the arena? If you need some help, try using a ruler or some compasses.

ANSWER: PAGE *108*

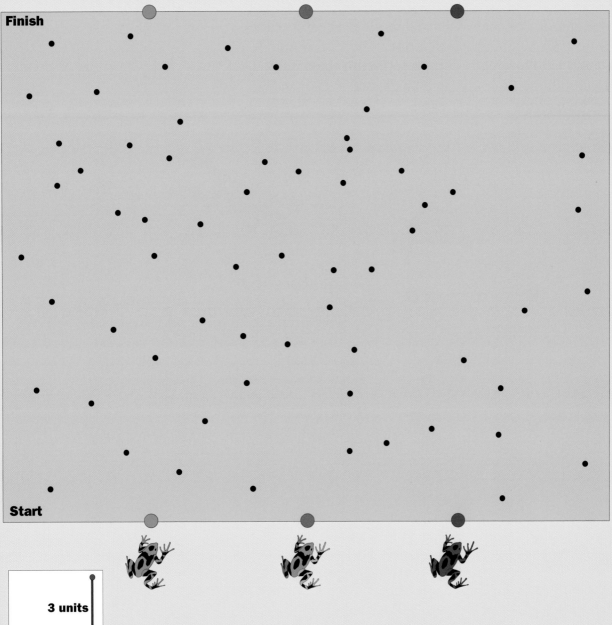

Finish

Start

3 units

2 units

1 unit

Frog 1   2   3

▲ HOPPING MAD

Three frogs (red, green, and blue) crossed the field from start to finish in a series of equal-distance jumps. The red frog's jumps were always 1 unit, the green frog's jumps 2 units, and that of the blue frog 3 units (the distances are shown in the table on the left). They had to cross the field in zigzagging jumps because of many obstacles. Their jumps are shown among many random points. Can you work out how many jumps each frog needed to reach its destination?

ANSWER: PAGE 108

In relation to its size, the grasshopper has the greatest jumping ability of any animal. Amazingly, grasshoppers feel no pain because their nervous system is decentralized and has no cerebral cortex. So even if these puzzles drive you hopping mad, you'll be pleased to hear that at least the insect won't be stressed.

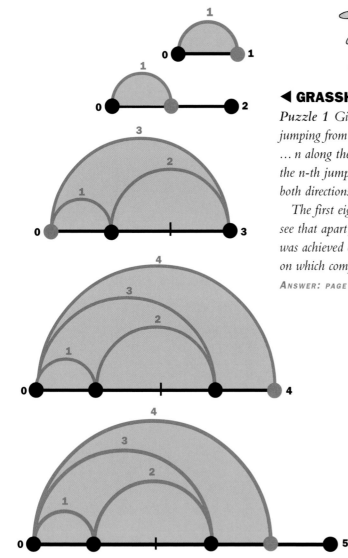

◄ GRASSHOPPING

Puzzle 1 *Given a line of integral length "n," the object is to start jumping from point 0, in successive jumps of consecutive lengths: 1-2-3-... n along the line, so as to make as many jumps as possible and finish the n-th jump at the end of the line, at point "n." Jumps are allowed in both directions along the line.*

The first eight games are demonstrated (from n = 1 to n = 8). We can see that apart from the trivial n = 1 line, a complete jump ending at "n" was achieved only on line n = 4. Can you find out the next two lengths on which complete jumps ending at the end of the line can be achieved?

ANSWER: PAGE 109

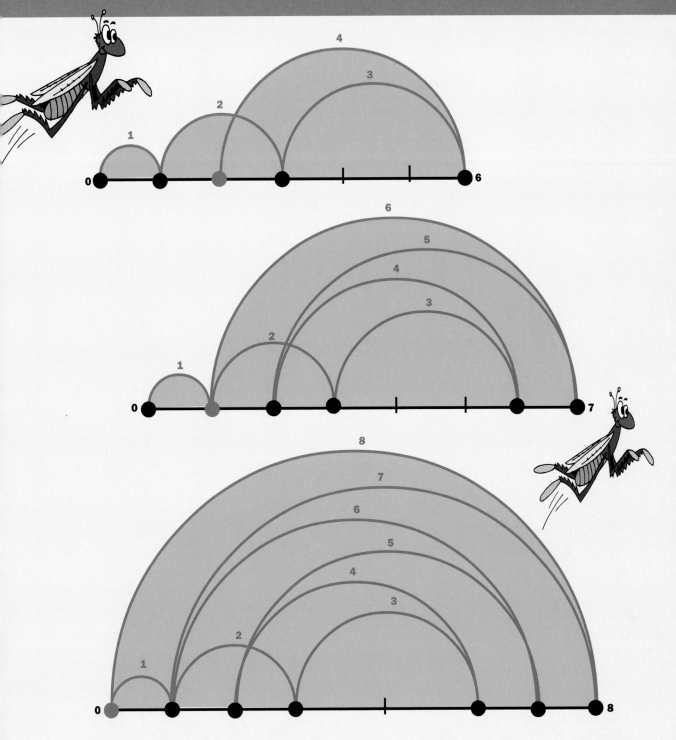

Puzzle 2 For some lengths, more than one sequence of moves is possible. For n = 6, there are two possibilities, both ending in four jumps. Can you find the other possibility?

ANSWER: PAGE 109

▼ GRASSHOPPING

Puzzle 3 Following the rules given on page 36, can you find out whether the lines $n = 16$ and $n = 20$ have solutions?

ANSWER: PAGES 109–110

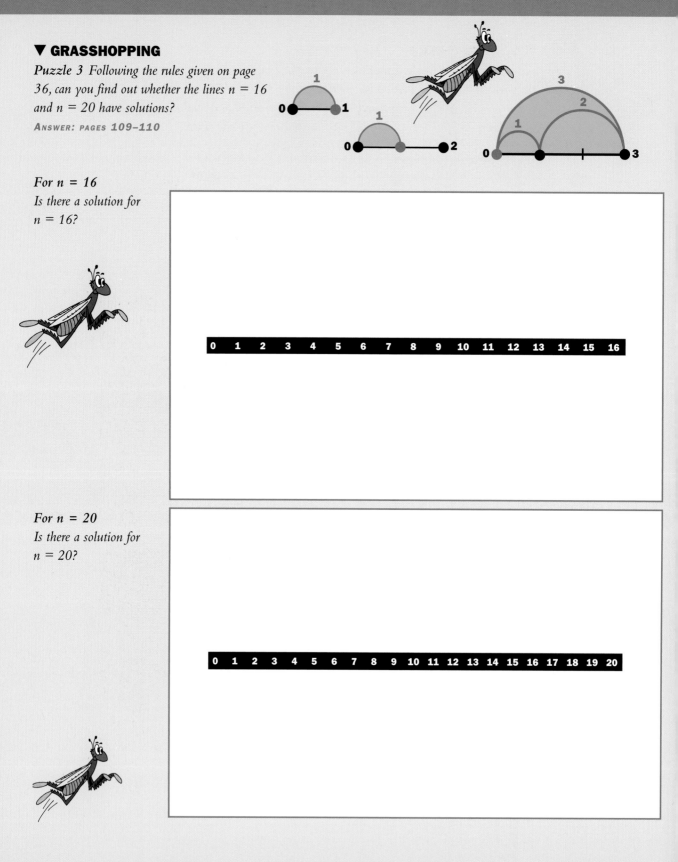

For n = 16
Is there a solution for $n = 16$?

| 0 | 1 | 2 | 3 | 4 | 5 | 6 | 7 | 8 | 9 | 10 | 11 | 12 | 13 | 14 | 15 | 16 |

For n = 20
Is there a solution for $n = 20$?

| 0 | 1 | 2 | 3 | 4 | 5 | 6 | 7 | 8 | 9 | 10 | 11 | 12 | 13 | 14 | 15 | 16 | 17 | 18 | 19 | 20 |

▼ LEAP FOR HOME

How many different paths is it possible for the grasshopper to take from the top of the upper left block to the top of the lower right block? The grasshopper can jump only one block down or one block right at a time. On its jumping way there are black holes that must be avoided. Clue: Page 17's puzzle may provide a hint to the method.

ANSWER: PAGE **110**

Hands up all those who remember everything they learned in math class! Try this puzzle, but don't be too dismayed if you find that a bit of geometry review is in order.

▼ SHAPE UP

Which of the following are true and which are false? If you think it depends, state what conditions are necessary for the statement to be true. ANSWER: PAGE 111

1. *All squares are rectangles.*

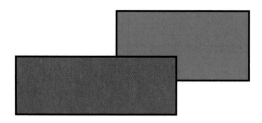

2. *A triangle can have two obtuse angles.*

3. *The diagonals of a rectangle cross at right angles.*

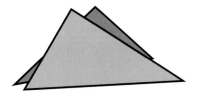

4. *When you double the side of a square you double the area.*

5. *A parallelogram has rotational symmetry of order 4.*

6. *The largest side of a triangle is shorter than the sum of the lengths of the other two sides.*

7. *If two rectangles both have the same area, they must also have the same perimeter.*

8. *A rhombus is a parallelogram.*

9. *Four infinitely long straight lines cross at six different points.*

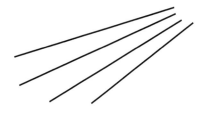

The mathematician John Horton Conway created a simulation called the Game of Life in which a grid of squares contained a number of cells simulating biological concepts. Overcrowded squares die off, as do barren squares, but those with an optimum population promote new growths—similar to how animal populations work.

▼ THE GAME OF LIFE

Usually, life simulation games (or, to give them their proper title, "cellular automata") are played on a 2-D grid. These normally involve using a computer program.

However, you can investigate cellular automata on a straight line quite easily. On the grid shown below, we're going to give birth to a red cell (value 1) on the right end, and a blue cell (value 2) just to its left. All the other cells have a value of 0. This initial setting on the top line is called Generation 1. What happens for the next line down (Generation 2) and all subsequent turns depends on these rules:

Rule 1 (birth): If a cell is in state 0 but its neighbors currently add up to 2 or more, its state in the next generation will be 2. Otherwise it will remain at 0.

Rule 2 (decay): If a cell is in state 2 and either neighbor is in state 0, its

state in the next generation will be 1. Otherwise it will remain at 2.

Rule 3 (death): If a cell is in state 1, its state in the next generation will be 0.

Using this set of three rules, can you continue the pattern for the next 13 generations? Once you've worked that out, try varying the starting conditions and have another go with the same rules. You can make up your own rules too!

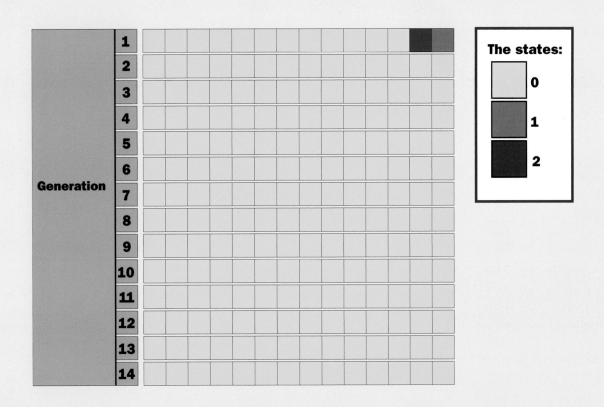

ANSWER: PAGE 111

The next time you go to Mali, remember this piece of etiquette: A woman will only shake hands with a man if he offers his hand first, with the left hand touching the other person's elbow. If that doesn't get you into knots, nothing will. See how you get on with these knotty handshake problems.

▲ GOLDEN HANDSHAKE

At the board meeting there were 17 members, each of whom is supposed to shake hands with every other person. But four members did not shake hands with each other. How many handshakes did happen?

ANSWER: PAGE *111*

▶ A SHAKER'S DOZEN

Twelve friends meet and shake hands with each other. How many handshakes were there altogether in this exchange of greetings?

ANSWER: PAGE 111

◀ LET'S SHAKE ON IT

My wife and I invited four married couples to our housewarming party. Just before everyone left, I asked everyone else how many people they had shaken hands with. I received the following replies: 8, 7, 6, 5, 4, 3, 2, 1, and 0. Given that no one shook hands with his or her own spouse and no pair of people shook hands more than once, can you tell how many times my wife must have shaken hands with a guest?

ANSWER: PAGE 112

Here's a mind-boggling statistic for you. Hawaii's main island has an area of 4,028 square miles. Since a square mile is 1,760 yards in each direction, that's 4,028 × 1,760 × 1,760 square yards. That multiplies up to such a large number that you could easily fit the entire population of the world onto Hawaii … even if they're lying down! How's that for a magic trick?

◄ T-HAT'S MAGIC!

The magician put four yellow, four green, and four red eggs in each hat. He called a person from the audience, blindfolded him, then asked him to transfer five eggs from hat 1 to hat 2.

The magician then asked the audience to tell the blindfolded assistant how many eggs have to be returned to hat 1 to ensure that in hat 1 there will be at least three eggs of each of the three colors. What is the answer?

ANSWER: PAGE 112

▲ ISLAND HOPPING

In how many distinct ways can you join the six islands shown above by bridges so that each island can be reached from every other island, and so that three of the islands have three bridges leading from them, two of the islands have two bridges leading from them, and one of the islands has only one bridge leading from it? Solutions that are 180-degree rotations (or mirror images) of another answer are not counted as different.

ANSWER: PAGE 113

These famous problems are sometimes called Monty Hall problems, named for Monty Hall, an American game show host who hosted "Let's Make a Deal." He was known for tempting people to give up their prizes to gamble for a mystery gift behind a door.

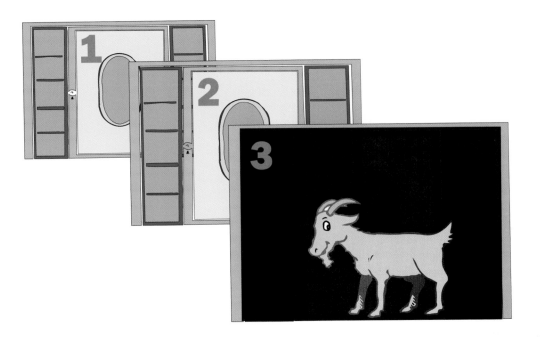

▶ MONTY HALL PROBLEM 1

You have been selected to participate in a game show that offers you the chance to win a luxury automobile. The automobile is behind one of the three doors. A goat is behind each of the other two. You choose a door that stays closed. The host (who knows where the automobile is) does what he always does: he opens an unpicked door to show a goat (there's always at least one unpicked door with a goat behind it to open). He then offers you the choice: To switch or not, that is the question?

Answer: page **113**

▼ MONTY HALL PROBLEM 2

For those who are still skeptical and are not yet convinced of the best method, play a number of problem 1 games with or without switching to check the validity of the answer.

Or, even better, we are providing a different version of the problem with ten doors, which should eliminate the mental blocks tens of thousands of people have when first encountering this problem.

As before, behind one of the doors is the luxury automobile and behind the other nine are goats.

You are allowed to choose one door that stays closed. The host always opens eight doors, behind all of which are goats. He leaves one door closed, and you again have the choice to switch. Will you? What are your chances of winning the automobile if you don't switch and stay with your initial choice? What are your chances of winning if you switch?

ANSWER: PAGE 113

Golomb rulers were introduced in 1952 by W.C. Babcock. They are named for Solomon W. Golomb, a professor of Mathematics and Electrical Engineering at the University of Southern California, who extensively analyzed the concept.

❋ Golomb rulers: perfect and optimal Golomb rulers

A Golomb ruler is a ruler constructed in such a way that no two pair of marks on it can measure the same distance. The markers on a Golomb ruler must be placed at integer multiples of a fixed spacing. The markers must be placed so as to achieve as many distinct measures of distances between two markers as possible, with a given number of markers.

In order to achieve this, the markers must be placed very efficiently, avoiding redundant distances between markers. In a perfect Golomb ruler of length n, all the distances from 1 – 2 – 3 – ... n can be measured exactly once, while in an optimum Golomb ruler (or the shortest Golomb ruler possible for a given number of marks), the condition remains that no two pair of marks can measure the same distance, but the ruler may not have all consecutive distances from zero to the ruler's length. (The distances correspond to what Golomb calls the "best" numbering of complete graphs for more than four points.)

A one-unit length ruler with two markers is "perfect" but trivial, that is, it can measure its one possible distance in only one way.

A two-unit length ruler with three markers is not "perfect," since it can measure a one-unit distance in two ways.

A three-unit length ruler with three markers is, in fact, the first "perfect" ruler. We can conclude so far that there are perfect rulers of length one and length three, but not of length two (see illustrations).

Finding and proving optimum Golomb rulers becomes more and more difficult as the number of marks on each ruler increases. Today optimum Golomb rulers up to 23 marks are known, and the search is on for the 24-mark and 25-mark rulers.

The Golomb rulers problem is considered to be one of the most beautiful problems in recreational mathematics. But such rulers are also needed in a variety of scientific and technical disciplines as well and are at the forefront of mathematical research, proving the relevance of recreational problems to pure math.

Golomb rulers provide a general spacing principle applied in astronomy (placement of antennas), X-ray sensing devices (placement of sensors), and in many other fields. Professor Golomb's further contributions are in the fields of combinatorial analysis, number theory, coding theory, and communication, all in addition to his major contributions in the fields of recreational mathematics, games, and puzzles.

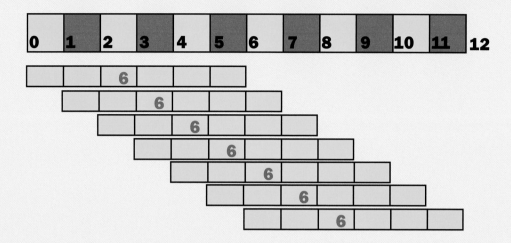

▲ GOLOMB RULER: INTRODUCTION

Is this a Golomb ruler? The 12-unit length ruler shown above with 13 markers (at every unit) allows us to measure any integer distance between 1 and 12 units.

From a mathematical viewpoint this is not a very economical arrangement. Using 13 markers to measure lengths from 1 to 12 is not very efficient.

For example, we can measure a length of 6 units in 7 different ways, and a 1-unit length in 12 ways. It obviously isn't a Golomb ruler. So the question is, can we reduce such redundancies and create Golomb rulers for different numbers of markers, optimal or perfect?

This 1-unit length ruler with 2 markers is "perfect" but considered trivial; that is, it can measure its one possible distance in only one way.

The 2-unit length ruler above with 3 markers is not "perfect," since it can measure a 1-unit distance in two ways.

A 3-unit length ruler with 3 markers is the first "perfect" ruler.

This demonstrates that there are perfect rulers of length 1 and length 3, but not of length 2, using 2 and 3 markers respectively. Can you find the next perfect Golomb ruler? Try the puzzles on the next two pages.

▼ GOLOMB RULER: 4 MARKERS

Placing the 4 markers this way would not produce a Golomb ruler, since three pairs of markers are separated by the same distance, measuring a 1-unit distance in three ways, and a 2-unit distance in two ways.

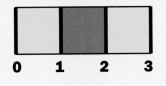

A better way of placing 4 markers produces a Golomb ruler. Remember: No two distances between markers are identical. The six distances that can be measured are shown here: 1, 2, 3, 4, 6, and 7.

Note that distance 5 cannot be measured and therefore this is not a perfect Golomb ruler.

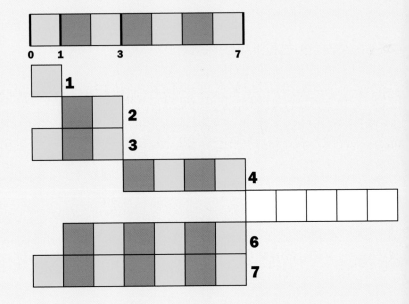

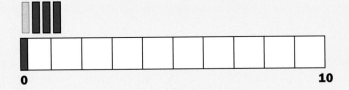

Can you do better and create a perfect Golomb ruler with 4 markers? What will its length be and where should the 4 markers be placed? The first marker has been placed for you, but where should the next 3 be placed?

Hint: Not all the squares are used, so the length is less than 10.

ANSWER: PAGE **114**

▼ GOLOMB RULER: 5 MARKERS

Five markers are placed on a ruler of 11-unit length. It is a Golomb ruler since no two pairs of markers measure the same distance. But is the ruler perfect? Can you measure all consecutive distance from 1 to 11 between any 2 markers, each distance, in only one way? Or, is it the shortest optimum Golomb ruler for 5 markers?

ANSWER: PAGE 114

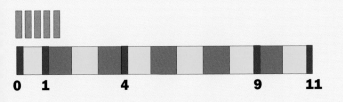

0 1 4 9 11

▼ GOLOMB RULER: 6 MARKERS

On a ruler of 17 units of length, it's not possible to place 6 markers so that you can measure every length from 1 to 17 units. However, see if you can place the markers so that all but two of these distances can be measured. Remember, each distance must have a unique way of measurement.

ANSWER: PAGE 115

0 17

▼ DISSECTED GOLOMB RULERS

If we changed the minimal ruler principle by allowing the ruler to be cut along the markers as shown on this 11-length ruler, and create the distances by rearranging the pieces, could we obtain a perfect ruler?

ANSWER: PAGE 115

0 1 4 9 11

7 markers

0 25

▼ GOLOMB RULER: 7 MARKERS

Can you place 7 markers on the 25-unit length ruler above, so as to be able to measure as many as possible consecutive distances from 1 to 25 unit lengths between 2 markers, each distance in one way only?

Or, in other words, can you create an optimum or shortest Golomb ruler? In the example above, 2 markers have already been placed, one at the beginning and the other at the end.

The divisions are there to help you to place the markers, but the distances can be measured only between 2 markers. In the ruler below, I have placed the 7 markers as shown. But is it the shortest or best Golomb ruler for 7 markers?

It can be seen that distances 10, 16, 17, and 24 cannot be measured. Can you do better?

ANSWER: PAGE 116

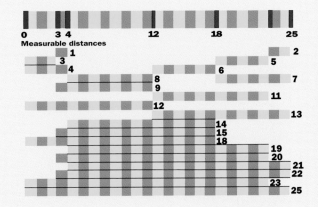

0 3 4 12 18 25
Measurable distances
1
3
4
8
9
12
2
5
6
7
11
13
14
15
18
19
20
21
22
23
25

Here's a thought—if two of each animal went onto the ark, does that mean there were one too many snails? Many species of snails are hermaphrodites and can fertilize themselves. While you're pondering that, we challenge you to tackle this menagerie of puzzles.

▲ BIRDS ON A WIRE

Imagine a very long wire on which a multitude of birds are randomly distributed, each looking one way or another at its nearest neighbor. No bird can see beyond its nearest neighbor. Can you guess how many birds will be observed by one, two, or none of its immediate neighbors if there is an infinite number of birds on the wire? In the example above, only 72 birds were randomly distributed.

ANSWER: PAGE 117

▼ ZOO-LOGICAL

The 7-by-5 unit rectangular area of the zoo has been divided into eight fenced regions to house the animals as shown.

Can you work out the size of the eight areas?

ANSWER: PAGE **117**

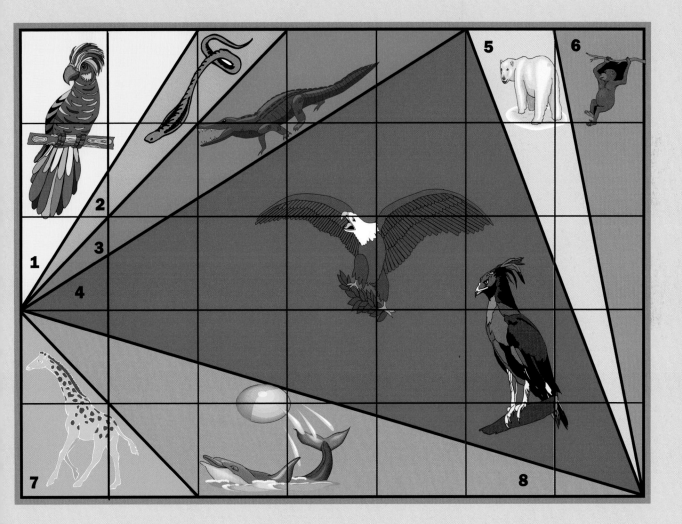

Any structure will necessarily contain an orderly sub-structure. Ramsey theory aims to figure out just how complex a structure must be in order to guarantee a certain substructure.

❊ Ramsey theory

The mathematicians Frank Ramsey and Paul Erdös stated that "complete disorder is an impossibility."

Stargazers have experienced the validity of Ramsey theory by finding patterns in the sky. Given a large enough number of stars, a pattern will be produced—from a perfect rectangle to the Big Dipper and others.

Although Frank Ramsey made considerable contributions to economics and philosophy, he is remembered more for his brilliance as a mathematician. The Englishman's best-known work was in set theory; indeed, a branch of that field now carries his name—quite an accomplishment for a man who died in 1930 at the age of 27!

The appearance of disorder is really a matter of scale: A mathematical structure can be found if you look widely enough. Ramsey wanted to find the smallest set of objects which would guarantee that some of those objects would share certain properties. For example, the smallest number of people

that will always include two people of the same sex is three. If there are only two, you might have a man and a woman; since the third person would have to be be either a man or a woman, adding him or her guarantees at least two of one sex.

Or take this question: Can a complete graph have its edges colored using only two colors so that no three edges of the same color form a triangle? Ramsey proved some general theorems on this question, but instances with four, five, or six nodes are simple enough to analyze using pencil and paper. The famous Party Puzzle on the page opposite is based on Ramsey's work.

To appreciate how useful graphs are for solving such problems, imagine listing all possible combinations of acquaintanceship among six people—a total of 32,768! And what if you then had to check whether each combination included a certain relationship! You'd be there forever.

A more advanced Ramsey problem would be to imagine a party where there must be a foursome in which everyone is a mutual friend or everyone is a mutual stranger. How large must the party be? Ramsey's work demonstrated that 18 guests are necessary. If you draw a complete graph with 18 nodes, no matter how you color the lines using two colors, you will inevitably create a quadrilateral formed by connecting four points (people) in one of the colors. The party size required to ensure at least one fivesome of mutual friends or strangers is still unknown. The answer lies between 43 and 49!

The beauty of Ramsey theory is its simplicity and the fact that it can be understood intuitively. Ramsey and Erdös contributed to designs of better communication and transportation networks by applying the theory, which, although still in its early stage, offers enormous implications for the future of mathematics. Quite an accomplishment from just a simple puzzle!

◀ RAMSEY GAME

The 15 white lines in the hexagonal graph on the left can be colored in either of two colors—red or blue. Two players alternate coloring lines one by one, in either of the two colors. The first player forced to create a triangle of one of the colors, connecting three points of the graph, is the loser.

How many different triangles can be colored? How many lines can be colored before there is a winner? And lastly, which player has an advantage?

ANSWER: PAGE 117

PARTY PUZZLE FOR SIX

In a party made up of you and your friends, any two of you are either mutual friends or mutual strangers. Can you invite five friends to a party and avoid having groups of three who are all mutual strangers or mutual friends?

We can simplify the problem representing the six people (you and your five friends) using a complete graph of six points. A complete graph is one in which each point is connected to every other point. In such a graph, we can clearly distinguish between all possible triplets, which form interconnected triangles, including interconnected pairs forming lines between points (persons).

If we decide that two people who know each other are connected by red lines in the graph, and two people who are strangers are connected by blue lines, we can solve the problem in the ingenious way devised by Paul Erdös, the famous Hungarian mathematician:

One by one, color the lines of the graph in either of two colors—red or blue. If you can choose your coloring so to avoid creating a triangle of one of the two colors formed by connecting three points, you have succeeded in avoiding having a group of three who are either mutual friends or mutual strangers.

The question is, can you achieve such a result in a group of six people?

ANSWER: PAGE 117

▲ RAMSEY 17-GON GAME: A THREE-COLORING GAME

Players alternate coloring a line using any of three colors until one of the players is forced to complete a solid color triangle, losing the game.

Can a game be finished in a draw, that is, ended without any player having created a solid color triangle? Again, the only way to find out is to have a go. (See the sample game right for how not to do it!)

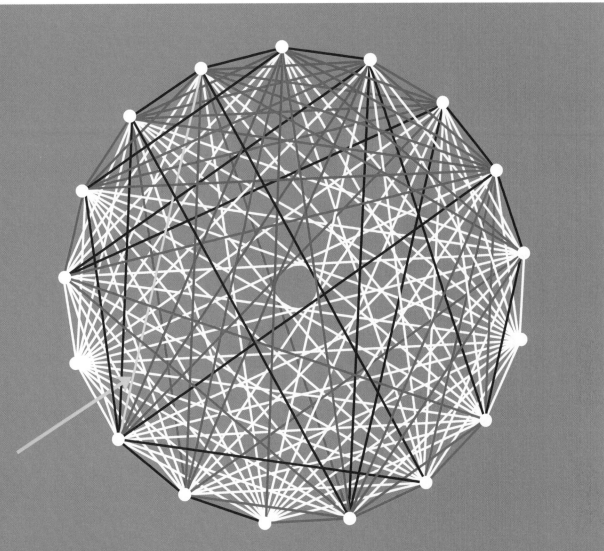

▲ SAMPLE GAME

In this sample game, we are guaranteed a winner. Why? Take a look at the arrowed line that we've colored in yellow. Even if every other white line can be validly colored, one player will be forced to color this final line in red, green, or blue. No matter which color is chosen, a triangle in the same solid color will be formed.

▲ RAMSEY 18-GON

Each line between two outer points of the 18-gon is colored in succession in either of two colors—red or blue.

Can you color the whole graph so that no four points are connected by lines forming either a completely red or completely blue quadrilateral, like the one shown in the sample game right? Have fun trying!

Answer: page 117

▲ HIDDEN POLYGONS

How many regular polygons and stars can you count in the circle?

ANSWER: PAGE 117

Sometimes conceptual blocks are of one's own creation, while others may stem from deliberately confusing and misleading directions. Inventors of puzzles and illusions exploit conceptual blocks to lead suggestible minds up blind alleys.

▶ A WHOLE LOTTO NUMBERS

How long will it take you to find in succession all the numbers from 1 to 90 in the image on page 61?

Counting aloud from 1 to 90 shouldn't take longer than a minute or two, but you will discover it won't be easy to find in succession all the numbers from 1 to 90 in our puzzle. Some numbers will be quite elusive, and your perceptual powers will be put to quite a test. It will probably take you longer than 15 minutes on the first attempt. Anything less than that is noteworthy.

The more times you do it the more improvement you should notice. Make a note of your scores. Time yourself again after attempting the other puzzles in this book. Did your score improve? It should have. Your visual perception will certainly improve as you work on the diversity of puzzles and problems within these pages.

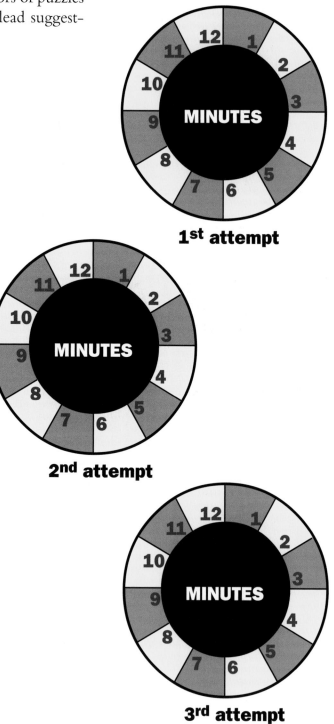

1st attempt

2nd attempt

3rd attempt

In major cities with mass transport systems, such as London and New York, there are fanatics who try to travel around every station in the quickest time possible. In a way, they are doing a real-time version of a well-known category of puzzles that was studied by the mathematician Euler (pronounced "oiler").

❊ Euler's problem

The Swiss mathematician Leonhard Euler lived from 1707 to 1783. At that time the town of Königsberg had seven bridges (see the illustration below), and it is said that the people of the town had never been able to solve the following problem: Is it possible to go out for a walk, cross each bridge once only, and return home?

Euler solved the problem by replacing it with something that looked simpler. Given an equivalent figure made of lines joining points, is it possible to traverse the figure in one continuous path, without taking pencil from paper and without going over any line twice?

Euler showed that for this to be possible there would have to be at most two places where an odd number of lines meet. And as a return to the start is a requirement, there would have to be no places where an odd number of lines meet. The reasoning is quite simple, once seen: A continuous journey will enter each junction exactly as often as it leaves—except at the start and finish.

The problem of the Königsberg bridges can therefore be solved by noting that it is equivalent to traversing such a network of lines, which has four junctions with an odd number of lines.

So it therefore follows that no solution can exist.

Euler's problem is really one of topology, a branch of mathematics which deals with properties of figures that are preserved by continuous deformations. Two networks are topologically equivalent if one can be distorted to give the other. If a network can be traversed by a single curve, so can any topologically equivalent network.

Another topic arising from Euler's work is graph theory, the study of networks formed by lines connecting points. Not bad for one recreational math puzzle!

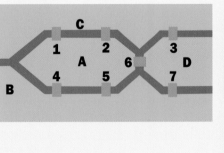

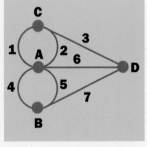

LEONHARD EULER (1707–1783)

Euler studied at the University of Basel, aiming to become a Protestant minister like his father, but his love of mathematics led him to change his studies. He was helped much by Johann Bernoulli, a friend of his family.

Euler's first job was at the mathematical-physical division of the St. Petersburg Academy of Science, where he was surrounded by a group of eminent scientists and mathematicians, including Daniel Bernoulli and Christian Goldbach. He was later appointed to the senior chair of mathematics.

The father of 13 children, Euler claimed that he made his greatest mathematical discoveries while holding a baby in his arms. His contribution to mathematics is enormous, in number theory, differential equations, calculus of variations, and other fields. His reputation was very high and in 1741 he was invited by Frederick the Great to the Academy of Science in Berlin, where he wrote: "I can do just what I wish—the King calls me his professor and I think I am the happiest man in the world." He spent 25 years in Berlin, writing about 400 articles before returning to St. Petersburg at the age of 59, where he became completely blind after an illness. Despite his blindness he produced nearly half of his total work there, relying on an extraordinary memory.

After his death, the Academy continued to publish Euler's works for 50 years or so.

❓ DID YOU KNOW?

Leonhard Euler wrote more mathematical research than anyone else in history.

❊ Possible-impossible traces

If a network has more than two odd vertices, it does not have an Euler path, nor an Euler circuit. If a network has two or fewer odd vertices, it has at least one Euler path.

Euler answered the questions of traversability by using the concept of valence and connectedness. The valence of a vertex in a graph is the number of edges meeting at that point. A graph is said to be connected if for each pair of its vertices there is at least one path of edges connecting the two vertices. A route that covers every edge in a graph only once and which starts and ends at the same vertex is called an Euler's circuit. If the route does not end at the starting vertex it is instead called an Euler's path.

There are two obvious questions to ask about Euler's circuits: 1. Is there a way to tell by calculation, and not by trial-and-error, if a particular graph has an Euler circuit? 2. Is there a method, other than trial-and-error, for finding an Euler circuit when one exists?

Now we can state Euler's famous theorem, giving a simple general answer to our problems: If a graph is connected and has all valences even, then it has an Euler circuit. The secret rule for tracing any puzzle of this kind is: You only have to check how many lines are going in or out from every intersection point. If there are more than two intersection points from which an odd number of lines emanate, the pattern is impossible to trace.

We may make an apparently innocuous change in Euler's problem and ask: When is it possible to find a route along the edges of a graph that visits each vertex once and only once in a loop or simple circuit? Such a route is called a Hamiltonian circuit—visiting each vertex in a graph. Note that in a Hamiltonian circuit some of the edges can be left untraversed. Though different, the concepts of Euler's and Hamiltonian circuits are similar in that both forbid reuse: the Euler circuit, reuse of edges; the Hamiltonian circuit, reuse of vertices. It is far more difficult to determine Hamiltonian circuits in a graph than Euler circuits.

E

uler's theorem is as follows:

1. A graph has an Eulerian circuit if and only if it is connected and all of its vertices are even;

2. A graph has an Eulerian path if and only if it is connected and has either no odd vertices or exactly two odd vertices. If two vertices are odd, then any Eulerian path must begin at one of the odd vertices and end at the other.

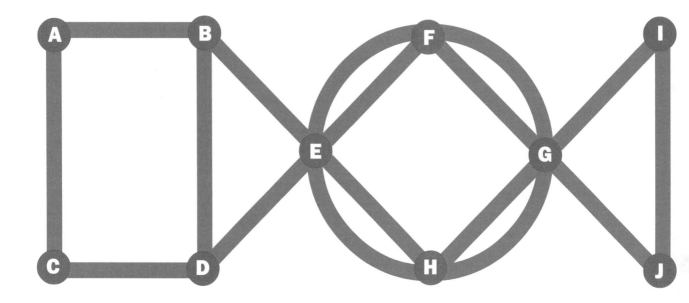

▲ EULERIAN PATHS

Take a look at the road network above. Starting at any point (A to J) that you choose, is it possible to drive over each road exactly once, thus creating an Eulerian path? Furthermore, is it possible to end up where you started from, thus forming an Eulerian circuit?

ANSWER: PAGE 118

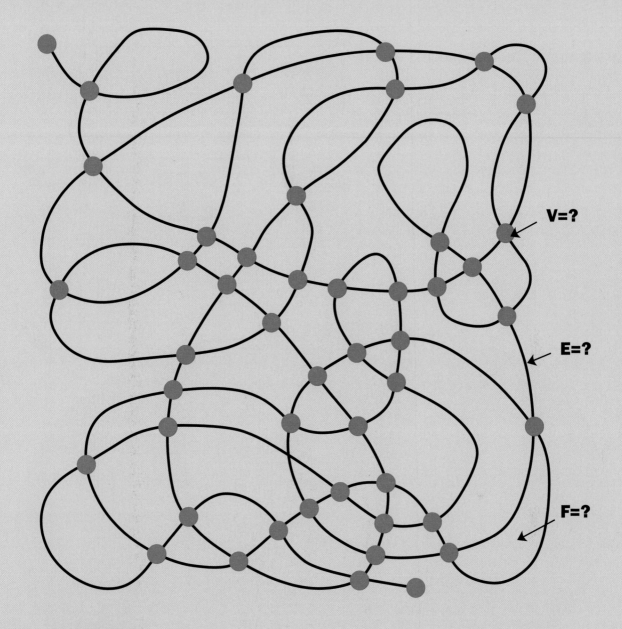

V=?

E=?

F=?

▲ DEVIOUS DOODLING

Did you know that doodles can be mathematically interesting? For example, try doodling a simple line that crosses itself a few times. Can you see a link between the number of vertices or intersection points (V), the number of edges or segments connecting two points (E), and the total number of regions including the outside (F)? Repeat the experiment a few times—do you find that the same rule holds up?

I have drawn a much more complex random doodle above. To make sure it was really random, I did it with my eyes closed! Can you use the rule you've discovered to determine V, E, and F for this diagram? You can count two of these amounts but you must calculate the third.

ANSWER: PAGE 118

▼ THE RIGHT CONNECTIONS

Three houses need three supplies: telephone, electricity, and water. So each house needs three connections. Can you draw in the connections (lines), to connect each house with each utility in such a way that no lines intersect:
1. in the plane (on the flat piece of paper)?
2. on the surface of a doughnut (torus)?

ANSWER: PAGE 119

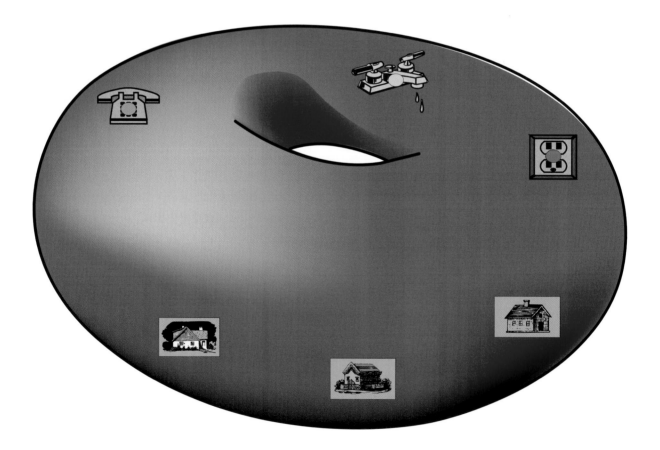

▶ CHINESE POSTMAN PROBLEM

In graphs for which Eulerian circuits are not possible because they have several odd vertices, some edges have to be crossed more than once. In such cases, the problem arises, how to do this most efficiently, so that the number of edges that are reused is minimal?

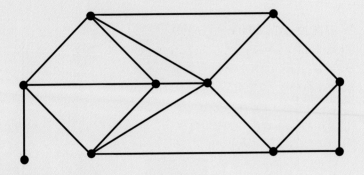

This can be done by altering the original graph to add new edges. This creates a new graph that has an Eulerian circuit—a process called Eulerization—which keeps the number of new edges to a mimimum. This problem is called the Chinese postman problem.

Guidelines for Eulerizing a graph:

1. It is permitted to add new edges only between vertices that have an edge between them.
2. All the odd vertices are circled in red.
3. For each pair of odd vertices, find the path with the fewest edges connecting them in the original graph, and then duplicate the edges along this path.

Note: Since there is always an even number of odd vertices, it is always possible to pair them. An example of this algorithm for Eulerizing a graph is shown below. Can you Eulerize the other diagram above?

ANSWER: PAGE 119

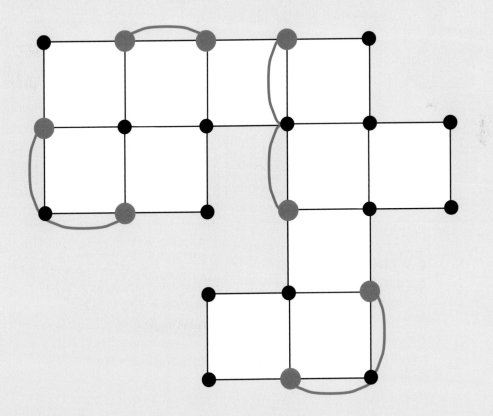

▲ DOING THE ROUNDS

The postman must deliver mail to each of the houses in the four-block neighborhood above, starting from one of the intersections. The streets on the outer edges have houses on one side, and in the inside on both sides. The route of the postman will take him once in the outer steets, and twice in the inner steets (once for each side). Can you find an Euler circuit here?

ANSWER: PAGE 119

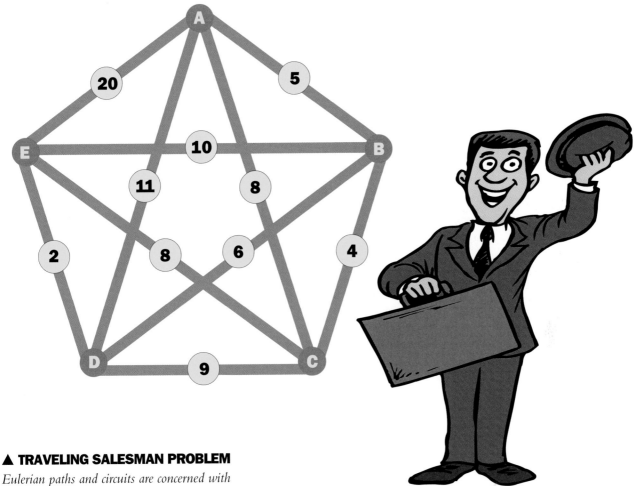

▲ TRAVELING SALESMAN PROBLEM

Eulerian paths and circuits are concerned with finding paths that cover every edge of a graph. Hamiltonian paths and circuits, on the other hand, deal with problems of visiting all of the vertices of a graph, without concern for whether or not all edges have been re-covered.

This type of problem was first studied by the Irish mathematician Sir William Rowan Hamilton, who was especially interested in finding a circuit that goes through every vertex exactly once and returns to the starting vertex. Today such routes are called Hamiltonian circuits. Those paths that don't return to the starting vertex after visiting every vertex are called Hamiltonian paths.

A related problem to Hamiltonian circuits is the traveling salesman problem: This problem involves finding a Hamiltonian circuit in a complete weighted graph for which the sum of the weights of the edges is the minimum. For example, suppose a salesman living at location B had to visit locations A, C, D, and E before returning home. The distance between every set of towns is shown by the yellow numbers.

Can you help find the most efficient route for the salesman so that he travels a circuit of the least possible total mileage?

ANSWER: PAGE 120

The British poet W.H. Auden once remarked to a TV journalist, "Your cameraman might enjoy himself, because my face looks like a wedding cake left out in the rain." Thankfully, the cakes on these pages have been kept nice and dry, which should help make it easier for you to cut them up.

◀ CHOP CHOP

At the birthday party three cakes are cut as shown below and divided between two groups, one group getting the red parts, and the other the yellow parts. Did both groups get an identical share of each cake?

Cake 1 is cut through the center, making six identical 60-degree angles, with the six pieces divided between two groups of three children. Cake 2 is cut through a point off-center, again making six identical 60-degree cuts and divided as before. Cake 3 is cut through the same point off-center as before, but making eight identical 45-degree angles, and divided this time between two groups of four children.

ANSWER: PAGE **121**

The cuts

cake 1   cake 2   cake 3

▼ CATERING FOR THE CHILDREN

At this birthday party three cakes have to be cut by vertical straight-line cuts into exactly 34 pieces and divided between 34 excited children.

Puzzle 1 What is the minimal number of straight line cuts needed to ensure that each child gets a (not necessarily identical) piece of cake? There is one condition: Each cake has to be cut by at least two cuts. Under this condition can each child get a piece of a cake?

Puzzle 2 What would be the minimal number of pieces and cuts needed if we change the problem so that it is now necessary to cut the three cakes into identical pieces (but again with straight-line cuts) so that each child gets a piece of cake?

ANSWER: PAGE **121**

cake 1   cake 2   cake 3

Shapes are not simply physical objects—they are also mathematical creations that can be described by numbers. And, like all numbers, they can be manipulated in different ways to get new results, a form of math that is known as geometric algebra. The use of algebra in geometry is now part of science, technology, and data analysis. Discover the facts on these pages, then turn to page 74 for a coordinated puzzle.

✳ From points to pictures

The concept of geometric algebra dates back to about 300 B.C., when Euclid used a form of it for proofs in his *Elements*. Geometry advanced little from the end of the Greek era.

The next big advance came from the French mathematician René Descartes, whose revolutionary treatise *A Discourse on Method* was published in 1637. It forged a link between geometry and algebra by applying the methods of one discipline to the other. This is the basis of analytic geometry, ushering in an unprecedented expansion of mathematical knowledge.

In analytic geometry, straight lines, curves, and geometric figures are represented by numerical and algebraic expressions using a set of axes and coordinates. Descartes was lying ill in his bed one day when he observed a fly landing on his ceiling, which was made of square tiles. He realized that at any time he could fix the position of the fly by the tile on which it was landing. This observation was the spark that led to the development of the Cartesian coordinate system.

With this method, a pair of numbers giving the distances from two axes could represent a point. Cartesian coordinates use axes x and y at right angles with an origin where the axes cross. When writing coordinates such as (2,3), the first number represents the distance along the horizontal (x) axis, and the second shows the distance along the vertical axis (y).

Equations can be used to plot shapes, if they are plotted in a coordinate system, as graphs. Fermat was the first to demonstrate equations in this way. If an equation has two variables, the shape is two-dimensional; if it has three, the shape is three-dimensional. Cartesian coordinates can be used to analyze curves. They can also help to solve simultaneous equations (the points at which the equation lines cross give their numerical solutions).

The German artist Albrecht Dürer drew human beings and animals in a coordinate system and by altering the coordinates obtained not only different proportions but also different characters and images.

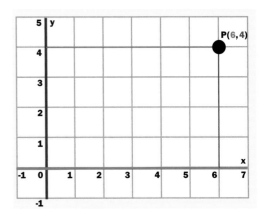

▲ COORDINATES ON A FLAT PLANE
Coordinates of a point P in a Cartesian system, with the values (x,y): x=6, y=4.

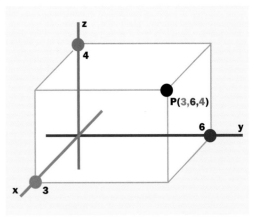

▲ COORDINATES IN 3-DIMENSIONAL SPACE
A third coordinate is added, z. Cartesian coordinates of a point P, having the values (x,y,z): x=3, y=6, z=4.

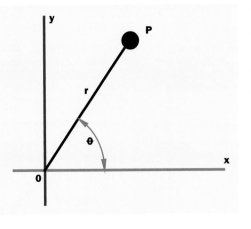

◄ POLAR COORDINATES

Instead of representing each point on the plane as an (x, y) coordinate, we could take a line of length r and rotate it by an angle Θ. Each point is uniquely identified by (r, Θ). Using the Pythagorean theorem and basic trigonometry, we can see how Cartesian and polar coordinates are related:

$$r = \sqrt{(x^2 + y^2)}$$
$$\Theta = \tan^{-1}\left(\frac{y}{x}\right)$$

In the example illustrated, r = 5, and Θ = 36.9 degrees.

$$x = r \cos \Theta$$
$$y = r \sin \Theta$$

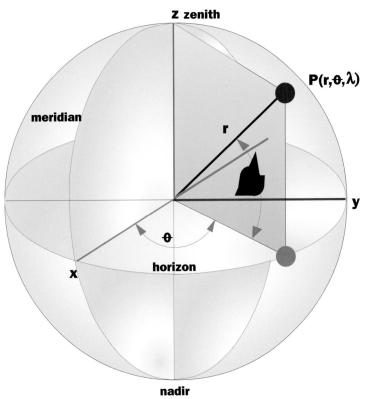

◄ SPHERICAL COORDINATES

Sometimes it is convenient to use polar coordinates in 3-D space, where the Pythagorean relationship gives us:

$$r^2 = x^2 + y^2 + z^2$$

where all the points of the same value of r form a sphere of radius r around the origin.

Astronomers can therefore use three simple numbers (r, Θ, λ) to define any point in the night sky (this is much easier to use than 3-D Cartesian coordinates).

Θ is the longitude or azimuth, that is, how far around the world we should look.

λ is the latitude or elevation, that is, how far we should tilt the telescope above the horizon.

r is the distance, that is, how deep into the sky we have to look.

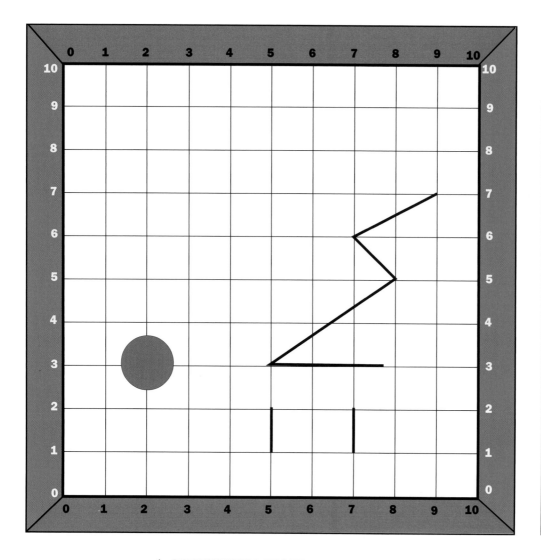

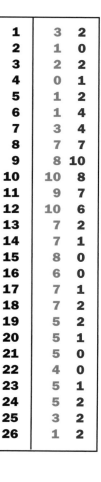

1	3	2
2	1	0
3	2	2
4	0	1
5	1	2
6	1	4
7	3	4
8	7	7
9	8	10
10	10	8
11	9	7
12	10	6
13	7	2
14	7	1
15	8	0
16	6	0
17	7	1
18	7	2
19	5	2
20	5	1
21	5	0
22	4	0
23	5	1
24	5	2
25	3	2
26	1	2

▲ COORDINATE CRAFT

A point on a two-dimensional plane of the paper can be determined as the intersection of two lines (coordinates).

Can you discover the picture that can be produced by interconnecting, in succession, the 24 points obtained by their coordinates?

ANSWER: PAGE 122

Many chess puzzles consider knight's tours or how many queens you can fit onto the board. But no one seems interested in the most important piece, the king. So let's fix that right now.

▶ ROYAL LINEAGE

In how many different ways can a chess king walk on a 3-by-3 chessboard from point 1 (lower left) to point 2 (top right)?

The only permitted moves from square to adjacent square are up, diagonal-right up, and right.

ANSWER: PAGE 122

▶ REGAL PROCESSION

In how many different ways can a chess king walk on the chessboard from point 1 (lower left) to point 2 (lower right)?

The only permitted moves from square to adjacent square are right, diagonal-right up, and diagonal-right down.

ANSWER: PAGE 122

It's funny how some numbers have special significance. In China, the number "four" sounds like "death" so you'll often find there's no 4th floor in buildings. And if you ever go on a date in Russia, buy your carnations in an odd number because even numbers of flowers are reserved for funerals!

▲ THINK OF A NUMBER

Think of a number from 1 to 10. Multiply that number by 9. If the number is a two-digit number, add the digits together. Now subtract 5. Determine which letter in the alphabet corresponds to the number you ended up with:

1	2	3	4	5	6	7	8	9	10	11	12	13	14	15	16	17	18	19	20	21	22	23	24	25	26
a	b	c	d	e	f	g	h	i	j	k	l	m	n	o	p	q	r	s	t	u	v	w	x	y	z

Think of a country that starts with that letter. Remember the last letter of the name of that country. Now think of the name of an animal that starts with that letter. Remember the last letter in the name of that animal. Think of the name of a fruit that starts with that letter. Compare your result with mine in the answer section.

Answer: page 123

▼ MAKE A MEAL OF THIS

We show below a piece of spaghetti, a pizza, and a potato cut by two straight cuts into three, four, and four pieces respectively. Here they represent a one-dimensional line, a two-dimensional plane, and a three-dimensional object.

How many pieces will result in each case if we use four cuts to get the maximum number of pieces? The number triangle is the key to the solutions, but how? Can you discover the secret rule behind the structure of the triangle that will give you the general solution?

ANSWER: PAGE 123

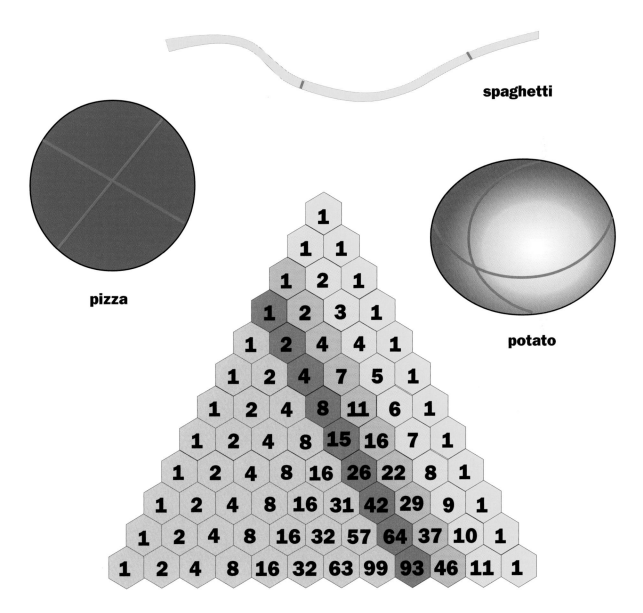

spaghetti

pizza

potato

Ever get the feeling you're going round in circles? Well, due to the Earth's rotation, you are. So be careful with the games on these pages—they could literally drive you round the bend.

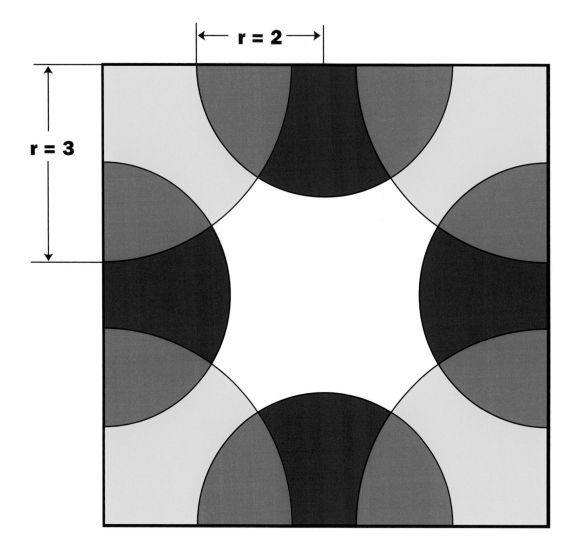

▲ CIRCULAR CONVERGENCE

Just by looking at the symmetrical pattern of overlapping identical quarter circles of radius 3 units, and identical semicircles of radius 2 units, can you tell what the relationship is between the yellow and blue areas?

ANSWER: PAGE **124**

▶ POLYGON BRACELET

These eight linked regular polygons will pull out into a circular shape to make a lucky charm bracelet.

In your mind's eye, can you work out what is special about the polygons that are opposite each other?

Can it really be a lucky charm?

ANSWER: PAGE **124**

Trial and error and a good dash of guesswork (plus a pinch of patience) may help you with the puzzles on these pages.

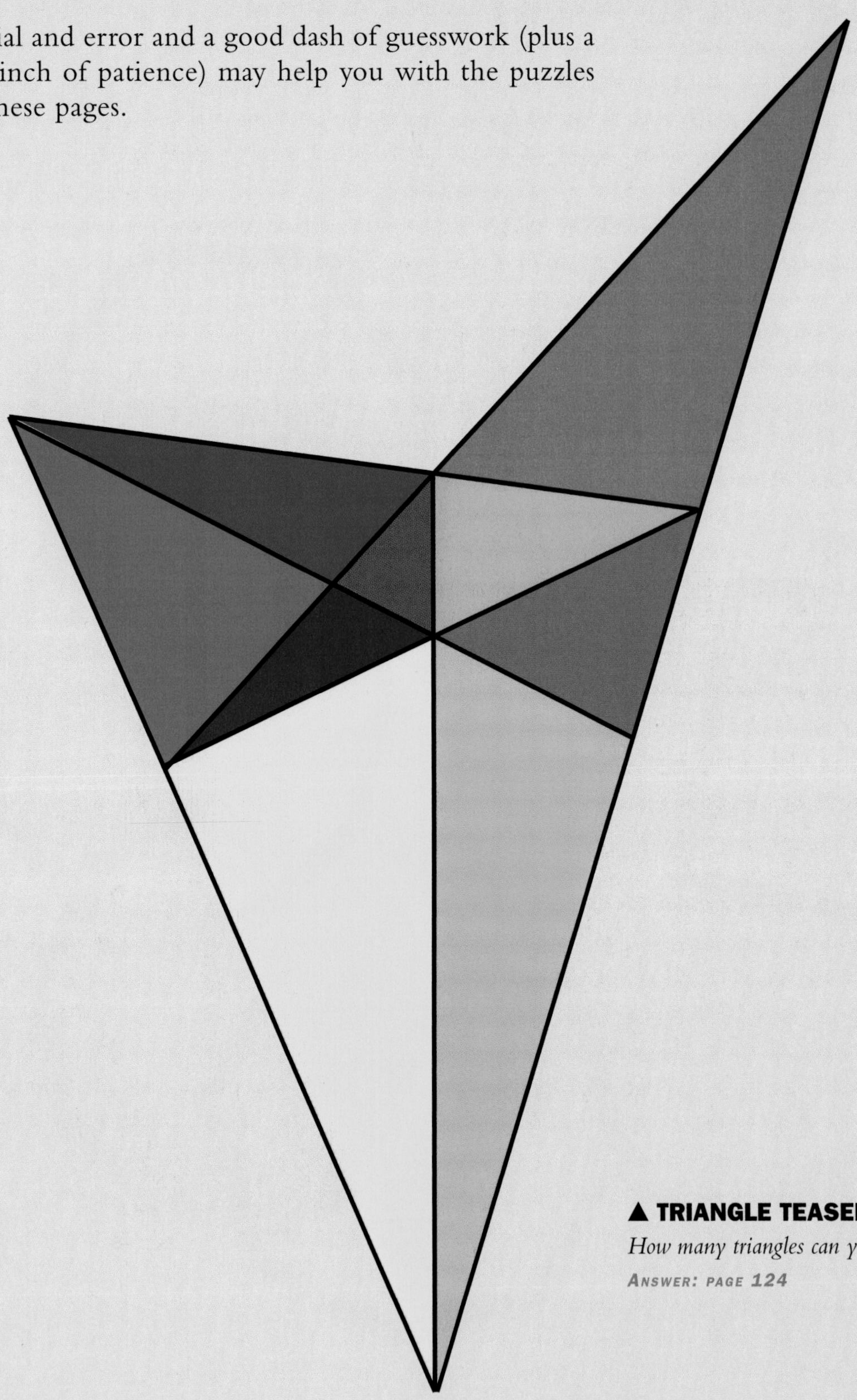

▲ TRIANGLE TEASER

How many triangles can you count?

ANSWER: PAGE **124**

▶ FLUID DYNAMICS

Colored water is poured into these two compartments as shown. Then the upper compartment is hermetically closed by its lid. The thin tube on the right is closed initially. What will happen when the right tube is opened?

ANSWER: PAGE 124

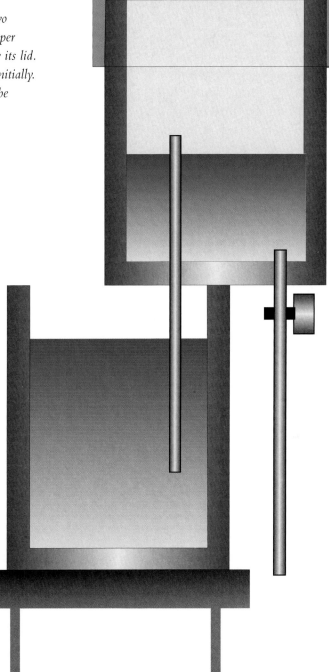

Minimal length circle problems are sometimes called "Ganymede circles," named for the third moon of Jupiter, one of the largest planetary satellites in our solar system.

▶ GANYMEDE CIRCLE 1

Earlier in this book, we met the concept of a Golomb ruler. A quick reminder: The aim was to place marks on a ruler so that, ideally, all the whole numbers from 1 up to the length of the ruler could be measured.

If we bend a Golomb ruler and join the ends together, we effectively get a Ganymede circle. So the aim here is to place three marker discs on the points around the circle so that it is possible to measure all the distances from 1 to 7 units by measuring from one marker to another. (For consistency, measure clockwise each time.)

Where should the three markers be placed?

ANSWER: PAGE **124**

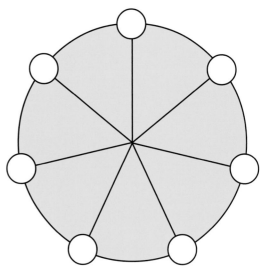

division 7

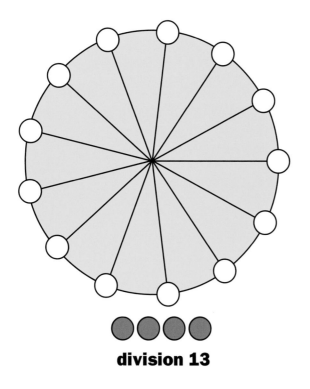

division 13

◀ GANYMEDE CIRCLE 2

The length of the circumference of the circle is divided into 13 parts. Place four marker points along the circumference so that every distance from 1 to 13 will correspond to a distance on the circle between two marker points.

ANSWER: PAGE **125**

▼ GANYMEDE CIRCLE 3

The length of the circumference of the circle is divided into 21 parts. Place five marker points along the circumference so that every distance from 1 to 21 will correspond to a distance on the circle between two marker points.

ANSWER: PAGE 125

division 21

▼ GANYMEDE CIRCLE 4

The circumference of the circle below is divided into 31 parts. Can you place six red marker points along selected points on the circumference so that every distance from 1 to 31 will correspond to a distance between two marked points?

ANSWER: PAGE 125

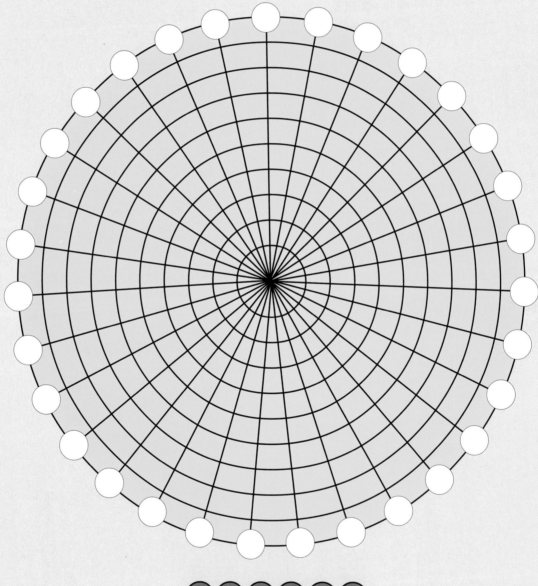

division 31

▲ COLORFUL COMBINATION LOCK

Imagine that these eight concentric rings can be rotated independently in any direction. Which one of the six colors will form four radial lines across the pattern?

ANSWER: PAGE 126

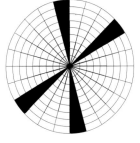

This is an original variation of the classical 4-by-4 magic square, with which you may be more familiar.

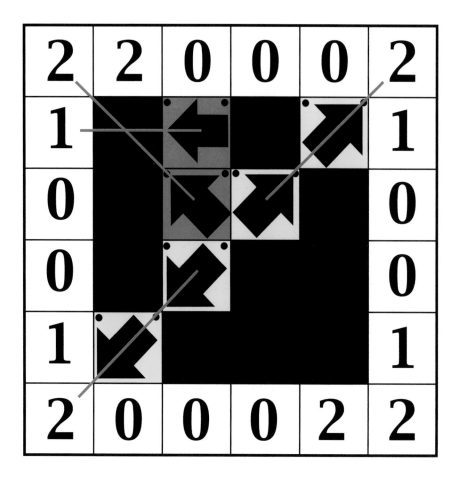

▲ MAGIC ARROWS: GAME 1

Sixteen arrow-blocks with arrows in different orientations can form a great number of different configurations. Some of those configurations form the basis of the puzzles on the next few pages, in which a set of number overlays are placed around a gameboard. The object of each puzzle is to re-create the related configuration of the 16 arrow blocks.

The 16 arrow blocks have to be placed in the gameboards with the two black points on top in such a way that the number of arrows pointing to a number on the cards will match the number on that card. Can you complete the game we've started above on the gameboard opposite? For each puzzle there may be different solutions other than the ones we've shown.

ANSWER: PAGE 126

▼ GAME 1

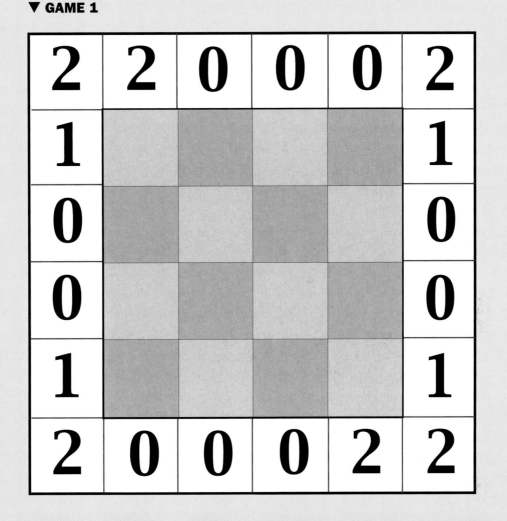

2	2	0	0	0	2
1					1
0					0
0					0
1					1
2	0	0	0	2	2

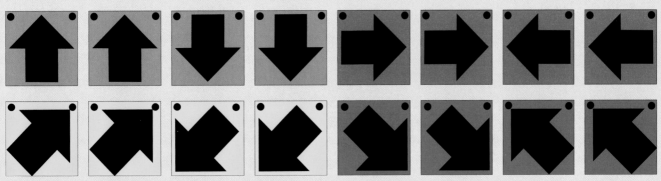

▼ MAGIC ARROWS: GAME 2

ANSWER: PAGE 126

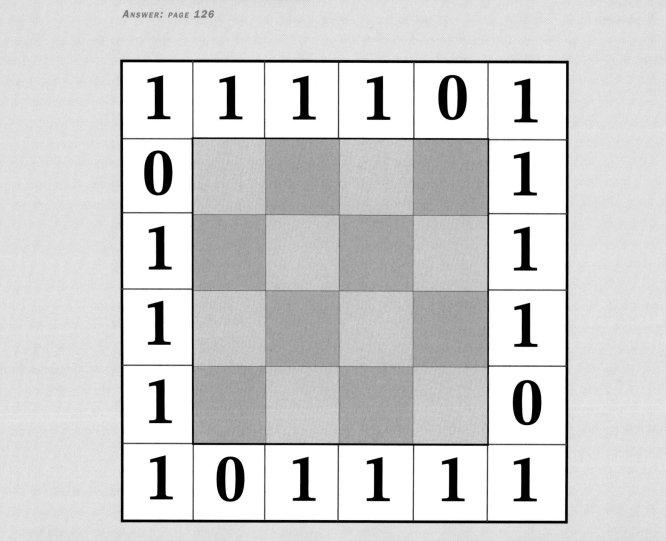

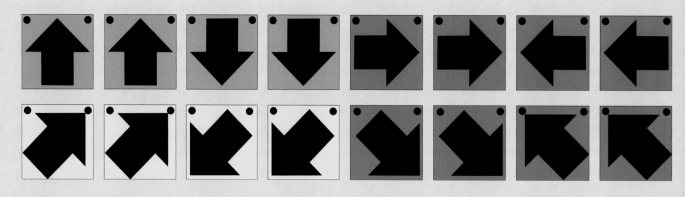

▼ MAGIC ARROWS: GAME 3

ANSWER: PAGE 126

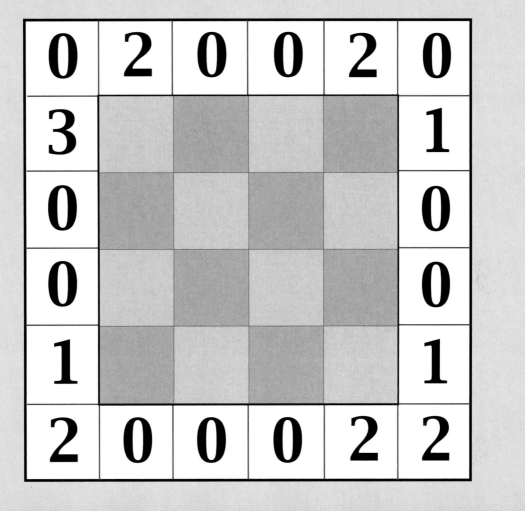

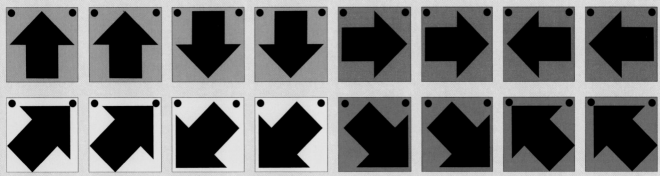

The four-leaf clover is universally accepted as a symbol of good luck. The legend goes that Eve carried a four-leaf clover from the Garden of Eden. See if you're lucky with the puzzles below.

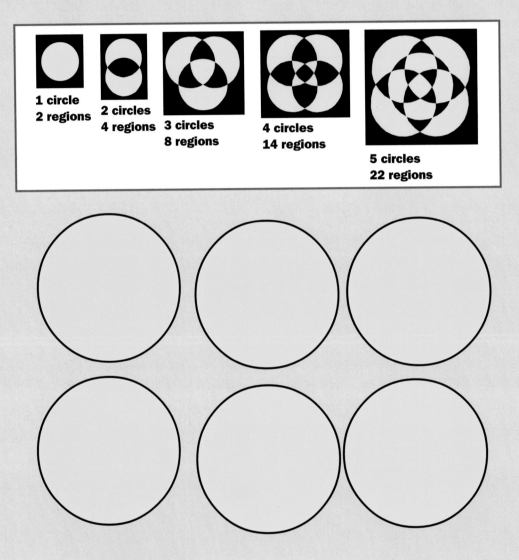

1 circle
2 regions

2 circles
4 regions

3 circles
8 regions

4 circles
14 regions

5 circles
22 regions

▲ IN YOUR AREA

Into how many regions do "n" identical circles divide the plane if each two circles intersect in two points and no three of the circles pass through the same point? We have illustrated the solutions for up to five circles. Can you find the number of regions into which six circles divide the plane?

ANSWER: PAGE **127**

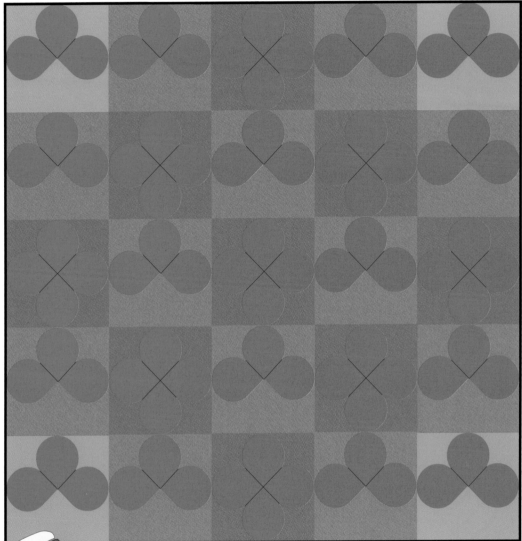

▲ CLOVER COVERAGE

The field is covered with clovers of three and four leaves.

The radius of their leaves is r = 1 unit length as shown.

Can you work out the proportion of the area covered by the clovers to the total area of the square field? (This puzzle was inspired by the ancient Sangaku-Japanese Temple Geometry tablets.)

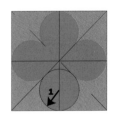

ANSWER: PAGE 127

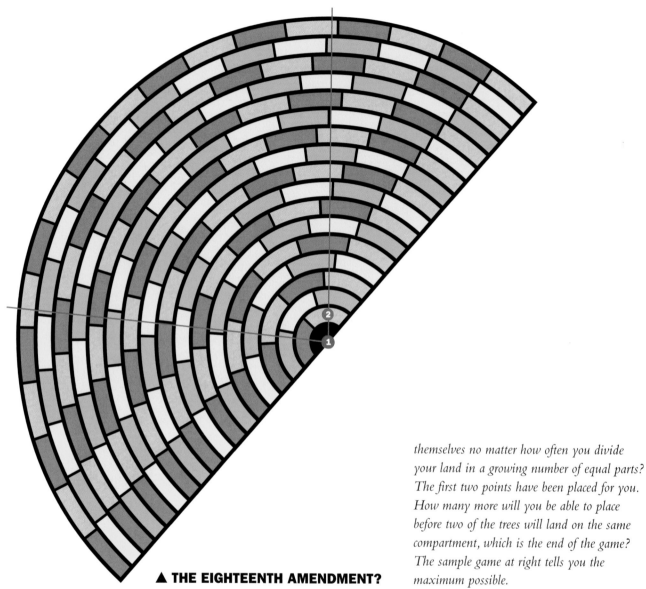

▲ THE EIGHTEENTH AMENDMENT?

Imagine you have a strip of semicircular land in which there is a tree, represented by a point that can be placed anywhere (1–center).

Dividing the land into two halves, you plant another tree in the second half (2). Then you decide to divide your land again, and plant another tree (3). And again. And again. Each time, the trees already planted turn out, luckily, to be in their own separate plots.

Can you be foresighted—and farsighted— enough to plant your trees where they will be by themselves no matter how often you divide your land in a growing number of equal parts? The first two points have been placed for you. How many more will you be able to place before two of the trees will land on the same compartment, which is the end of the game? The sample game at right tells you the maximum possible.

The strips of land are represented by parts of concentric semicircles. But remember: The increasing concentric semicircles are representing the same length of land divided into increasing number of equal divisions.

A challenging two-person game can also be played. Players alternate by placing their trees. The loser is the player who on his or her turn is forced to plant a tree in an area already occupied by a tree planted earlier.

ANSWER: PAGE **128**

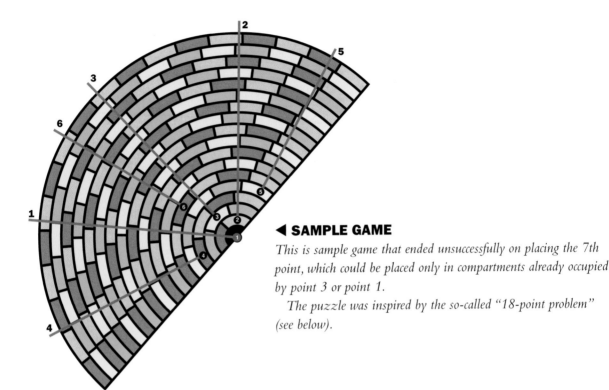

◄ SAMPLE GAME

This is sample game that ended unsuccessfully on placing the 7th point, which could be placed only in compartments already occupied by point 3 or point 1.

The puzzle was inspired by the so-called "18-point problem" (see below).

❋ The 18-point problem

Mathematicians sometimes invent seemingly simple, trivial-looking problems that prove much more difficult to solve than anyone dare think. One such conundrum is the 18-point problem, mentioned by Martin Gardner in his "Mathematical Games" column in *Scientific American* magazine.

The object is to distribute 18 points along a line segment according to simple rules. Line segments, of course, comprise a multitude of points—indeed, an infinite number of points are on a line. So you might imagine that with sufficient foresight,

one could place an infinite number of points on a line segment. That intuition, however, turns out to be wrong.

The rules of the game are quite simple: Place a point anywhere on the line segment. Now place a second point so that each of the two points lies on a different half of the line segment. Then place a third point so that each of the three points is in a different third of the line segment.

At this stage it becomes clear that the first two points cannot be just anywhere; the points must be placed carefully so that when the third point is

added, each will be in a different third of the line segment.

The game follows a predictable pattern—place the fourth point so that all are on different quarters, the fifth so that all are on separate fifths, and so on. You can proceed with this process as carefully as you wish, but it turns out, somewhat astonishingly, that you cannot go beyond 17 points. The 18th point will always violate the rules of the game. Even when you choose the locations of your points very carefully, placing 10 points counts as a good result.

Take the humble equilateral triangle, stick six of them together and, hey, presto—you have a regular hexagon. These two puzzles highlight interesting relationships between triangles and hexagons.

▶ ACCESS ALL AREAS

If a regular hexagon and an equilateral triangle have the same perimeter, what is the relationship between their areas?

ANSWER: PAGE **128**

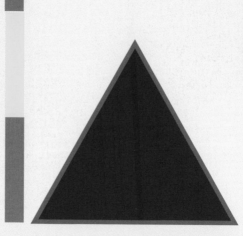

▼ TO THE MAX

Which of these six triangles on the sides of a
regular hexagon has the greatest area?

ANSWER: PAGE 128

And so here we are at the end. But maybe by now you're looking at math in a whole new light. So let's call this the end of a beginning rather than the beginning of the end.

▲ ONE FOR THE BIRDS

In a cage are red birds and blue birds. Each red bird sees as many red birds as there are blue birds. Each blue bird sees three times as many red birds as blue birds. How many red and blue birds are in the cage?

ANSWER: PAGE **128**

◄ THE SAFE SAFE

The safe has four dials, each with the 26 letters of the alphabet, and is opened by a four-letter code, in which each letter can be used only once. The order matters, so XBFG is different from GXBF. Do you think it would be easy for a thief to figure out the secret code? If it takes five seconds to try each possible combination, how long do you think it would take to try them all?

ANSWER: PAGE **128**

► WINDOW OF OPPORTUNITY

It took the maid too much time to clean the window so her employer ordered a new one, giving the builder exact instructions: The new window should give half the light, it should be a square, and it should be the same height and width as the present window. How did the builder solve the problem?

ANSWER: PAGE **128**

▼ INEBRIATED INSECT (page 6)

Here, the "net view" shows the cylindrical glass flattened out into a rectangle. The shortest path involves the ladybug crawling upward to meet the edge halfway, then taking a similar path back down inside the glass.

To calculate the length of the path: Circumference of glass = π × diameter, so the rectangle's width = 3.14 units × 4 = 12.6 units (approximately). The outside ladybug will travel half of this distance (6.3 units) horizontally, and a total of 5.5 units vertically. Via the Pythagorean theorem, we can therefore deduce the total distance traveled = square root of (6.3 squared + 5.5 squared) = 8.4 units to one decimal place.

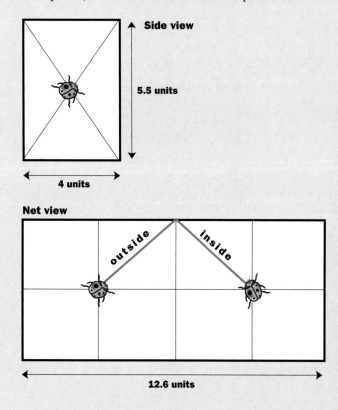

▼ ▶ THE SHORT AND LONG OF IT (page 7)

The longest path is shown below (lengths used are indicated with a yellow box). It measures 320 units in total. Very often simple analog gadgets can solve complex mathematical problems. Jos Wennmacker of the Netherlands devised a simple gadget that can solve our problem and even more complex problems of this kind in no time.

Wennmacker creates an analog model of the graph by knotting together pieces of string in exact scale (or connecting the pieces of string to small rings or eyelets). This is demonstrated on page 99. The result is obtained by two simple operations. Pick up the string structure at any node (point) and let it hang freely. Pick up again at the lowest node and hang it again, and you have the longest path. As simple as that! The selected node for the first step is shown with a red circle around it. The longest route is shown by the squares on the map (below).

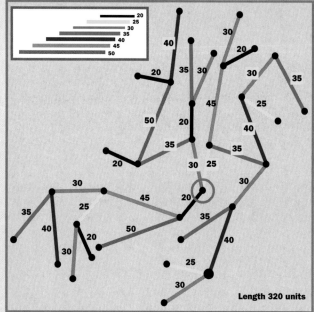

WENNMACKER'S GADGET

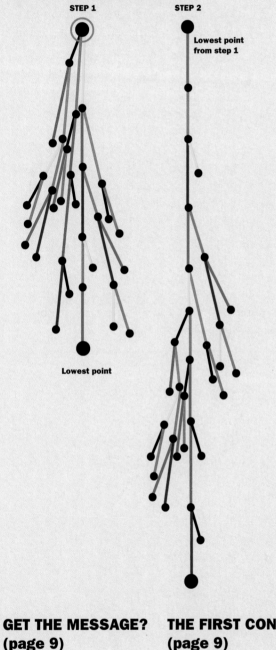

STEP 1

STEP 2

Lowest point
from step 1

Lowest point

▼ BIANGLES (page 13)

Eight examples are shown here. Others are possible.

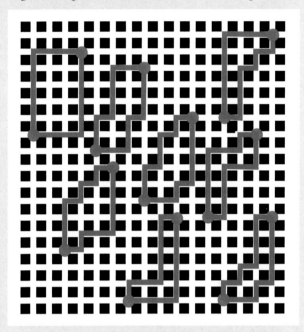

▼ TAXI TRIANGULATION (page 14)

The smallest area of a triangle with sides 6, 8, and 14 units in taxicab geometry is 13 block units, as shown here in three examples.

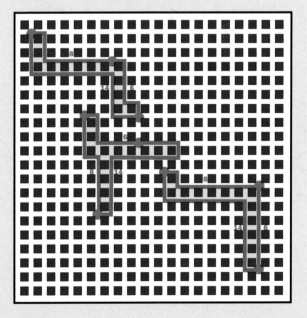

GET THE MESSAGE? (page 9)

1 + 2 = 3
2 + 3 = 5
5 − 2 = 3

THE FIRST CONTACT (page 9)

1 + 2 = 3 (true)
2 + 3 = 5 (true)
5 − 2 = 3 (true)
3 + 2 = 4 (false)

A convincing proof that the message was understood.

▶ SQUARE ROUTE (page 15)

To find the squares that enclose the fewest blocks, we need to force the taxicab to drive as close to the city center as possible, while maintaining a route of exactly six blocks between each pair of points. The minimum squares possible are shown here (15, 12, and 27).

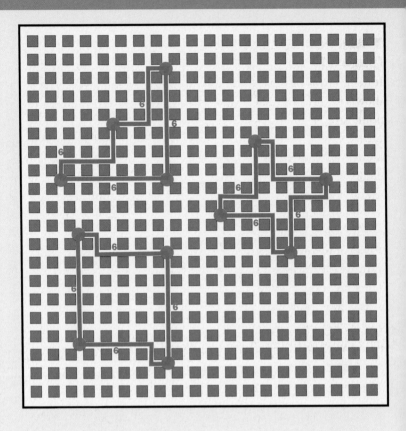

▶ GOING ROUND IN CIRCLES (page 16)

Surprisingly enough, in taxicab geometry circles are squares. The circle of 7 unit radius is shown in red and the intersecting circle of 5 unit radius is shown in green. The two circles intersect at 11 points as shown in blue.

Although Euclidean geometry states that any two intersecting circles can have at most two points in common, taxicab geometry allows circles to intersect at any number of points. The larger the squares ("circles"), the more points they can have in common.

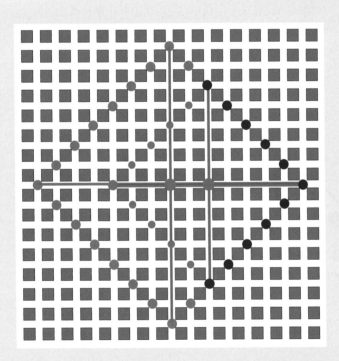

▼ ALL ROADS LEAD TO...
(page 17)

It helps to tilt the grid 45 degrees as shown for reasons to be explained. The point now at the top is "home."

Let's first consider why there may be more than one shortest route between two points in Gridlock City.

There is only one way to stay at home, so we put 1 in that position. There is exactly one way to go to each of the points in the next layer down, so mark 1s in those too. Now consider the third layer. The only way to get to its left-corner is by the end of the layer above. But the middle position can be reached in two ways: either by the position above and to

the left, or above and to the right. Then there is only one way to reach the right-hand end. So place the numbers 1, 2, 1 on the diagram. And so on.

By now you should have worked out the principle. The numbered triangle is the famous Pascal's triangle. In Pascal's triangle each number is the sum of the two numbers above it. This triangle has many applications in probability.

The two tan limiting routes are of 20 units length. There are many more routes, all inside the yellow square as shown. The final figure shows there are 184,756 possible routes.

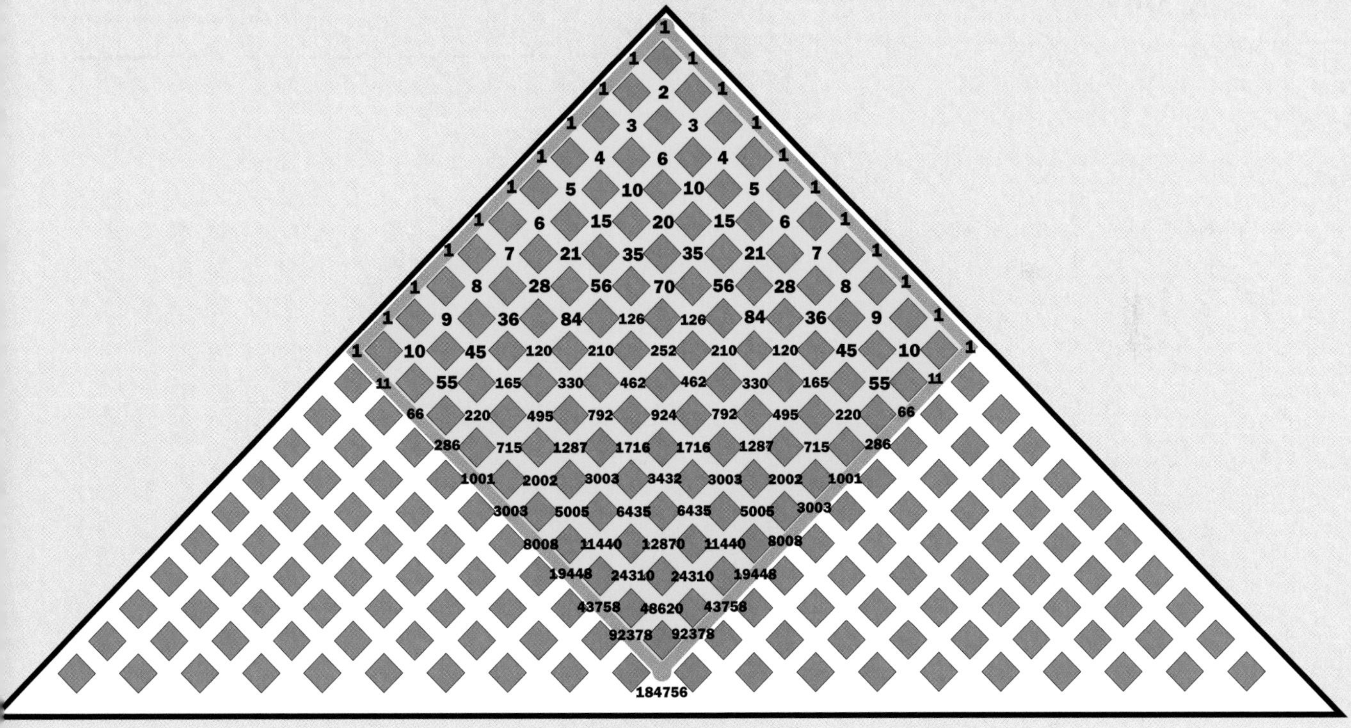

LOCAL LIQUOR (page 18)

The way to find the average "best" location is to average the x-coordinates first, and then the y-coordinates:

Average of x = (1+1+1+3+4+6+7+7+8+10+11+12)/12=5.9

Average of y = (1+1+2+3+4+4+5+7+7+9+9+12)/12=5.3

Therefore the best place will be (5.9, 5.3).

The nearest coordinate to this is at (6,5), which is point B.

▶ TREASURE HUNT (page 19)

In Euclidean geometry, the treasure would be at one of the two points of the circle at a distance of 7 units from your chosen point as shown (see answer to Going Round in Circles on page 100).

In Gridlock City you will need just one trial dig to pinpoint the treasure, since the treasure is hidden at one of the two intersection points of the two circles as circled here in red.

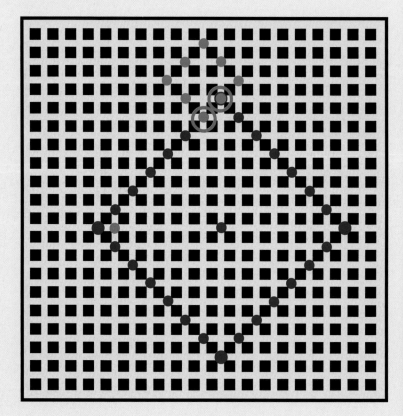

▶ CUBIC CRYPTOGRAM
(page 20)

The secret behind the spy's message is revealed here: "BETRAYED-LEAVE-GRIDLOCK-NOW!"

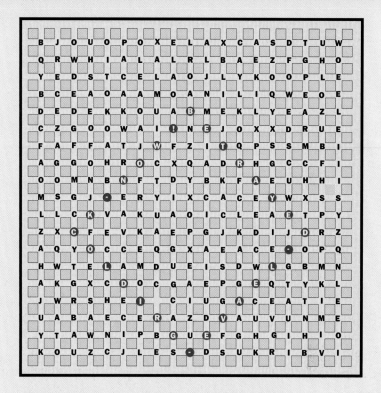

▶ DELIVER THE GOODS
(page 21)

The schematic diagram on the right shows the minimum distance between every combination of locations in Gridlock City. For example, the shortest taxicab distance between the red "3" and the red "5" is six blocks, so we label that line with a yellow 6.

The problem now becomes: Visit every red dot in a loop, collecting the smallest total of yellow numbers possible. Via a little trial and error, you'll find the shortest solution is 26 blocks long, as shown by the arrows on the diagram.

By analogy, the best route on the original diagram (on page 21) is to visit the locations in this order: 1, 5, 6, 2, 3, 4, and back to 1. As we've already seen, there are many different "shortest" routes you could take that follow this route.

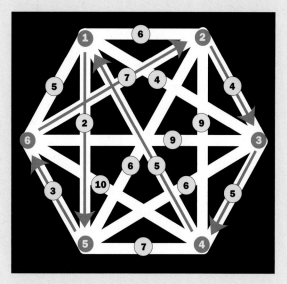

▶ **SCENIC TOUR (page 22)**

One of the many possible journeys is shown here.

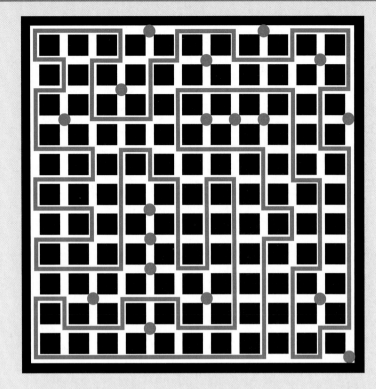

◀ **RIGHT THIS WAY (page 23)**

Some possible routes are shown on the left.

▼ AREA OF A CIRCLE (page 24)

The circle is first cut into sectors that, arranged as in the drawing below, form a rectangle.

The more sectors into which the circle is divided, the more the sectors resemble triangles, with the triangles becoming smaller and smaller and the figure approaching the form of a rectangle.

The height of each triangle is roughly the same as the radius of the circle (call this R). The circumference of a circle = 2 × π × R. Each color of triangle covers half of the circle's circumference, so the length of the rectangle is half this, that is, pi × R. Hence: area of circle = area of rectangle (due to our rearrangement) = height × width = R × (π × R) = π × (R squared), which is the famous formula we know today.

Note that this is really an approximation. The method actually works only if the base of each triangle ("a" in the original diagram) is infinitely small.

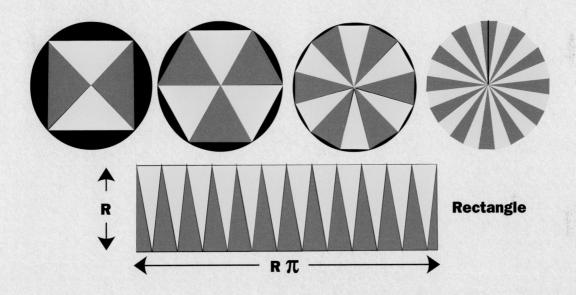

R

R π

Rectangle

CLOVER (page 25)

The two areas are identical.

4-WAY RACE (page 26)

Surprisingly, the ball on the cycloidal track will be the first to arrive. This is only one of the remarkable properties of the cycloidal curve.

A descending ball will roll on an inverted cycloid in a shorter time than it would roll on any other path, straight or curved, and this in spite of it being the longest path. It is quite amazing to think that the shortest path (the straight line) is not actually the quickest route.

The cycloid is therefore called the "curve of the quickest descent"—the brachistochrone.

Why is this? The ball descending along the cycloid reaches a higher speed in the early part of its descent and thus reaches the end first.

It is even more surprising that a descending ball will reach the bottom in the same time no matter from which point on the inverted cycloid it started.

Galileo discovered that the period of a pendulum depends only on its length, which is true only for small oscillations. By making the pendulum wrap around a cycloid, however, it becomes true for oscillations of any amplitude.

POLYGONAL PROOF (page 27)

The area under the polygonal arch generated by one vertex of a regular decagon rolling along a straight line is three times the area of the decagon (or any other regular n-gon—that is, polygon of "n" sides). You may have spotted that the colored triangles in the upper diagram can be rearranged into the three decagons shown underneath.

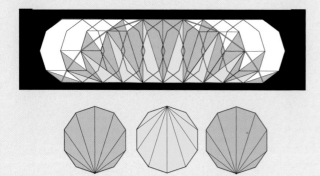

POLICE CHASE (pages 30–31)

The answers to our problems are related to the basic geometry of triangles.

Puzzle 1: The minimal triangle problem
In essence, we want to build three straight "roads" from each town to a point in the middle of the triangle. The puzzle asks us to minimize the total length of these roads. As it happens, this occurs when the roads meet at 120 degrees, as shown. The place where the roads meet, and thus where we should build our police station, is called the Fermat point.

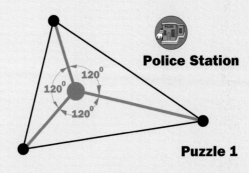

Police Station

Puzzle 1

Puzzle 2: Circumsizing a circle about a triangle The point that is nearest to any of the three towns can be found by constructing a circle that will pass through the three vertices of the triangle. The center of this circle is the point we are looking for.

To find it, we have to draw the three perpendicular bisectors on each side of the triangle. The point where they intersect is called the *circumcenter* of the triangle. Since the three towns are on the same circle they are at the same distance from the circumcenter.

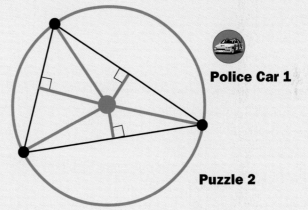

Police Car 1

Puzzle 2

Puzzle 3: Inscribing a circle in a triangle, or the angle bisection theorem The point nearest to any of the three main roads can be found by constructing a circle inside the triangle, which will be tangent to (touching) the three sides of the triangle. To find the center of this circle, called the *incenter* of the triangle, we have to bisect each angle and the point where the bisection lines meet is our point. Since the tangents are on the same circle they are at the same distance.

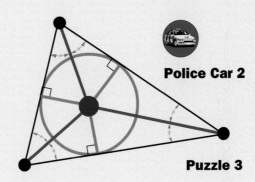

Police Car 2

Puzzle 3

▶ BRIGHT BLOSSOMS
(pages 32–33)

The water levels of the four vases will be as shown, visualizing the area relationships of the four shapes. Note how the green triangle has the same volume as the blue triangle because it has the same size of base and perpendicular height.

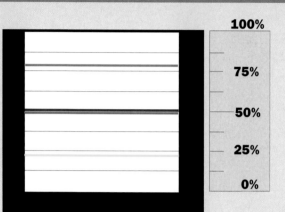

◀ FLEA CIRCUS (page 34)

Green takes 10 jumps, as shown, while red takes 22 jumps.

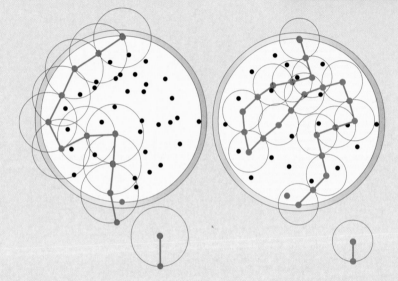

▶ HOPPING MAD (page 35)

Red frog—20 jumps
Green frog—8 jumps
Blue frog—4 jumps
Again, the puzzle can easily be solved using a compass or a ruler (as the ancient Greeks solved puzzles).

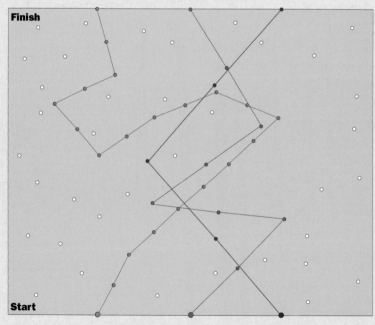

► **GRASSHOPPING**
(pages 36–37)

Puzzle 1 The next two complete jumps are n = 9, and n = 13, as shown.

The latter can be expressed as:
$$1 + 2 - 3 + 4 + 5 - 6 + 7 - 8 + 9 - 10 + 11 - 12 + 13 = 13.$$

Hint: As is quite often the case with some categories of puzzles, the solutions are facilitated if they are attacked from the objective backward, as shown in Puzzle 3 on page 38.

Puzzle 2 The other four-jump possibility for n = 6 is the same as that shown for n = 5 on page 36, but with a slightly longer base line.

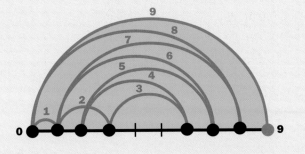

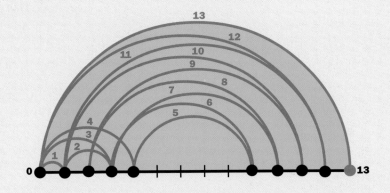

► **GRASSHOPPING**
(page 38)

Puzzle 3 It's more convenient to try to solve the puzzle working backward, with the assumption that there is a solution and that the longest jump (of 16 spaces) ends at point n = 16.

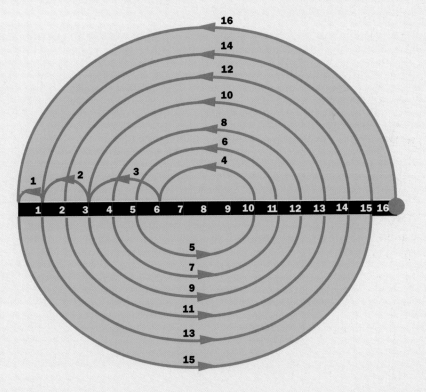

▶ GRASSHOPPING (page 38)

Puzzle 3 (contd.): The unique solution for
n = 20 is shown on the right.

The solution shows the method of
assuming that there is a solution and
going backward, starting with the longest
line n = 20.

This way it is easier to place the shorter
jumps to achieve a solution if there is one.

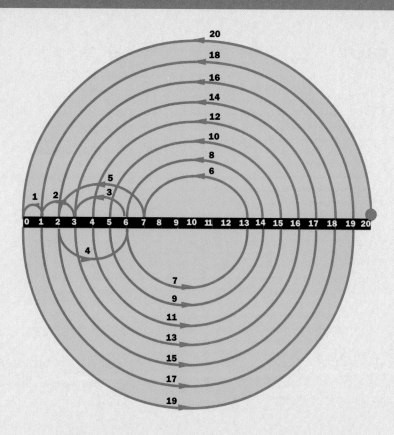

▶ LEAP FOR HOME (page 39)

Starting from the beginning, if we
consider the 2 × 2 square shown, then the
number of paths to the right block is "x"
and the number of paths to the lower
block is "y." Then the number of paths to
the lower right block is "x + y."

Following this pattern we can write in
each block the number of different paths
leading to the end block as shown.

SHAPE UP (page 40)

1. True
2. False
3. True only if the rectangle is a square
4. False (it quadruples)
5. True only if it is a square
6. True
7. False
8. True (it's a special case of a parallelogram—with all sides equal)
9. True if and only if no two are parallel, and no three lines pass through a single point, as shown below:

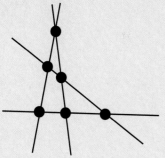

THE GAME OF LIFE (page 41)

The two-cell glider will "ripple" to the left infinitely.

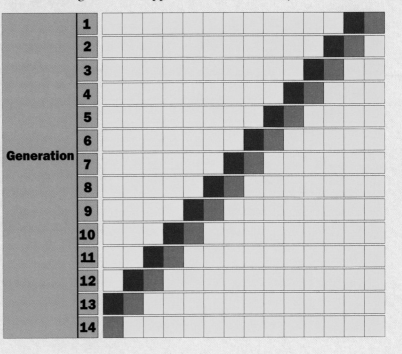

GOLDEN HANDSHAKE (page 42)

In general, "n" people shake hands with "n − 1" people (as one doesn't shake hands with oneself). Since two people share a handshake, this result must be halved to get the number of handshakes (H):

$$H = n \times (n - 1)/2 = (n^2 - n)/2$$

The 17 members of the board were supposed to shake hands with 16 people, which would amount to $(17^2 - 17)/2 = 136$ handshakes.

But four people did not shake hands, so there were $(4^2 - 4)/2 = 6$ handshakes fewer, that is, 130 handshakes.

A SHAKER'S DOZEN (page 43)

There were 66 handshakes, since $(12^2 - 12)/2 = 66$.
(See Golden Handshake on the left for a full explanation.)

▶ LET'S SHAKE ON IT (page 43)

A ten-point graph can help (to demonstrate the ten people involved). Assume A had the maximum number of eight handshakes. J was left without a handshake, so must be the wife of A because we are told that spouses did not shake hands.

B had seven handshakes, and I must be the wife of B (with one handshake).

C had six handshakes, and H must be the wife of C (with two handshakes).

D had five handshakes, and G must be the wife of D (with three).

E had four handshakes, and F must be the wife of E having also had four handshakes.

It follows that I must be E, and F (my wife) had four handshakes as well.

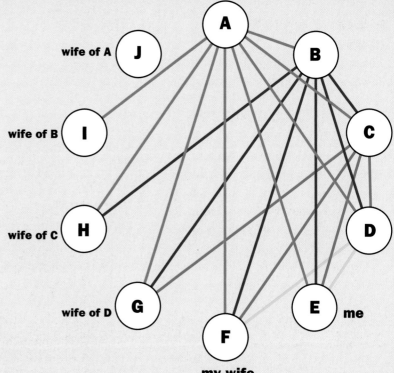

▶ T-HAT'S MAGIC! (page 44)

In move 1, the worst possible scenario is to transfer four eggs of one color plus one of another color.

So, in move 2, the worst scenario would be to transfer 9 + 3, or 12 eggs to satisfy the requirement to have at least 3 eggs of each color in hat 1. So 12 eggs is the answer the audience must give to ensure the magician gets the result he wants.

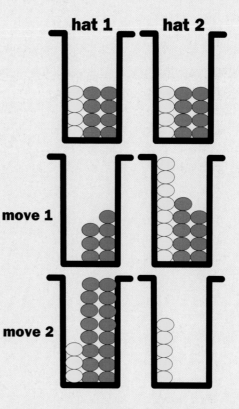

▶ ISLAND HOPPING
(page 45)

The six islands can be joined in 13 different ways as shown at right.

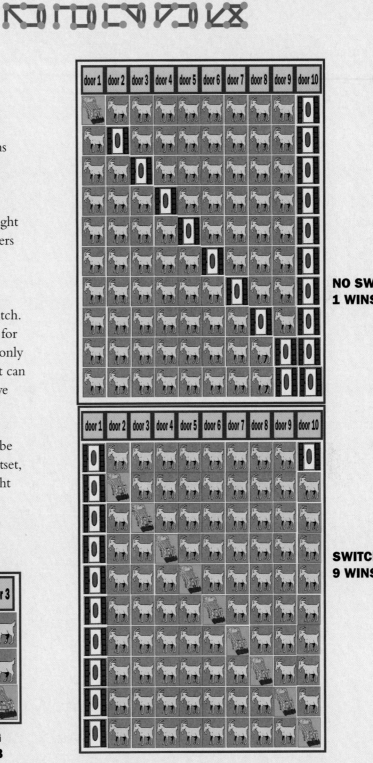

▶ MONTY HALL PROBLEMS
(pages 46–47)

Martin Gardner has presented several versions of the "game show" paradox, which still pops up in different versions and disguises, but the *Parade* magazine columnist Marilyn vos Savant is most famously associated with it. Her 1990 column on the subject provided the right answer but provoked thousands of letters of disbelief and accusation.

Why? Because the answer seems so wrong and counterintuitive.

The correct method is always to switch. Suppose we choose door 1. The chart for Problem 1 below shows how we win only 1 out of 3 times if we don't switch but can increase our chances to 2 out of 3 if we do switch.

It is easier to see in Problem 2. It's intuitive that the car is more likely to be behind one of doors 2 to 10 at the outset, and the chance that we picked the right door to begin with is only 10%. So switching is clearly the right strategy.

**NO SWITCHING
1 WINS IN 10**

**SWITCHING
9 WINS IN 10**

PROBLEM 1

**NO SWITCHING
WINS 1 IN 3**

**SWITCHING
WINS 2 IN 3**

▶ GOLOMB RULER: 4 MARKERS (page 50)

A 6-unit length ruler with 4 markers is "perfect."

The 4 markers on a ruler of length 6 enable you to measure all consecutive distances from 1 to 6 between 2 markers in one way only.

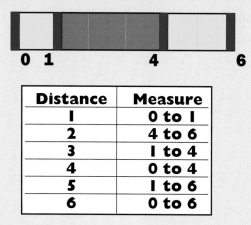

Distance	Measure
1	0 to 1
2	4 to 6
3	1 to 4
4	0 to 4
5	1 to 6
6	0 to 6

▶ GOLOMB RULER: 5 MARKERS (page 51)

No, it's not a perfect ruler. In fact, no perfect Golomb rulers exist that are longer than 6 units. For longer rulers, either some distances occur more than once or some distances cannot be measured at all.

In our 11-length ruler with 5 markers, the length 6 cannot be measured, although all the rest can be, as shown at right. It is the shortest optimum Golomb ruler for 5 markers.

Another 11-length ruler with one distance missing is shown below right. Here, the 10-unit length cannot be measured.

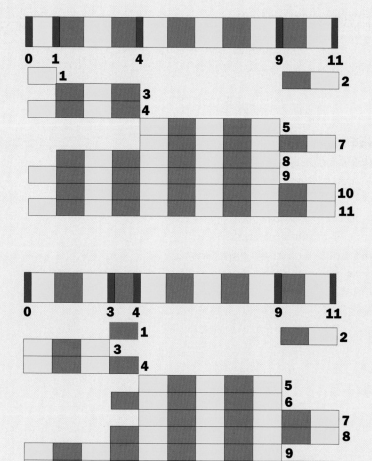

▼ GOLOMB RULER: 6 MARKERS (page 51)

With the markers placed as shown, all the distances can be measured, except distances 8 and 12. This demonstrates, therefore, that it is not a perfect ruler.

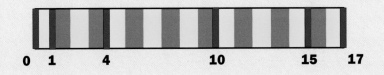

0   1   4   10   15   17

This is just one solution. Three other possible marker arrangements are:

(0, 1, 4, 10, 12, 17)—14 and 15 can't be measured

(0, 1, 8, 12, 14, 17)—10 and 15 can't be measured

(0, 1, 8, 11, 13, 17)—14 and 15 can't be measured

DISSECTED GOLOMB RULERS (page 51)

Allowing the dissection of the 11-length ruler and rearranging the pieces for creating the desired distances, we can measure all the distances from 1 to 11. By definition it is not a "perfect" ruler, however, because distance 3, for example, can be measured in more than one way.

Sidenote: By allowing dissection, rulers of length 7 and 15 can be made perfect.

▼ GOLOMB RULER:
7 MARKERS (page 51)

It's not possible to improve on the basic tenet—wherever the markers are placed, you cannot measure 4 of the distances from 1 to 25 units.

By way of example, another method of placing the markers is given here. The distances 11, 12, 16, and 20 cannot be measured. In a sense, this is considered to be a slightly better ruler because you can measure from 1 to 10 before hitting a problem, whereas for the one in the question you can only measure from 1 to 9 before getting stuck.

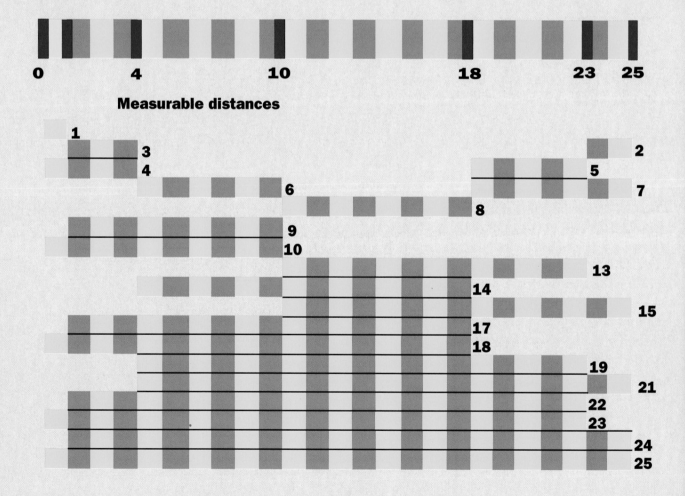

Measurable distances

BIRDS ON A WIRE (page 52)

With an infinite number of birds randomly distributed along the wire, 50% of the birds will be watched by one of its neighbors, and another 25% will be watched by two of its neighbors, while 25% will be left unwatched. The situation is similar to tossing a coin twice: 50% chance of one head, 25% chance of two heads, 25% chance of two tails.

ZOO-LOGICAL (page 53)

Area 1 = 3 square units
Area 2 = 1.5 square units
Area 3 = 3 square units
Area 4 = 15.5 square units
Area 5 = 2.5 square units
Area 6 = 2.5 square units
Area 7 = 2 square units
Area 8 = 5 square units
Total area = 35 square units

RAMSEY GAME (page 55)

No matter how you try to color the graph, you can't avoid forming a triangle of one of the colors, so a draw is impossible.

Twenty different triangles can be colored. A maximum of 14 lines can be colored before the 15th line forces a red or blue triangle. There is no best first move, and no best strategy found so far. The second player has an advantage.

PARTY PUZZLE FOR SIX (page 55)

Achieving the result needed would be like playing the Ramsey Game above with no outright winner. However, we have already seen that the 15th and final line always forces a result—one player must complete a red or blue triangle.

Therefore, our task is not actually possible—that is, you can always find a group of three people who are mutual friends or mutual strangers.

RAMSEY 18-GON (page 58)

No matter now many lines are colored, it is impossible to avoid forming a quadrilateral of one of the colors.

HIDDEN POLYGONS (page 59)

9 squares

8 squares

2 octagons

1 octagonal star

▶ EULERIAN PATHS (page 64)

We can conclude at the outset that the graph does not have an Eulerian circuit since it has two odd vertices (B and D).

However, we can start an Eulerian path at point B and finish at D, or vice versa. In creating the path we have to be careful not to make a move that would result in disconnecting the uncovered paths.

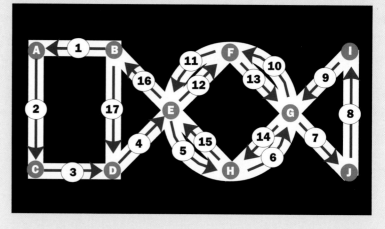

▶ DEVIOUS DOODLING (page 65)

In your experiment, you should have found that the number of vertices, edges, and regions (or "faces") are related by this formula known as the Euler Characteristic: $V + F = E + 2$

In our example, it seems easier to count the number of intersection points (there are 48) and the number of regions (there are 47). Hence, the number of edges can be calculated via:

$48 + 47 = E + 2$

$E = 93$

Something to investigate: Does this formula work for three-dimensional shapes (for example, a cube)?

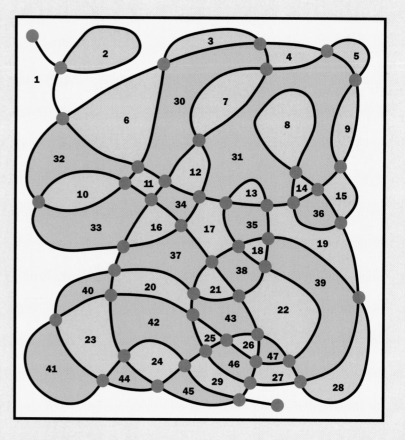

▶ THE RIGHT CONNECTIONS
(page 66)

In the plane, connections are impossible without at least one intersection or finding ways to cheat—for example, feeding one line through the middle of a house.

In the three-dimensional space, on a torus, a Möbius strip, or on the projective plane, however, it can be done without any intersections as shown (by looping lines underneath).

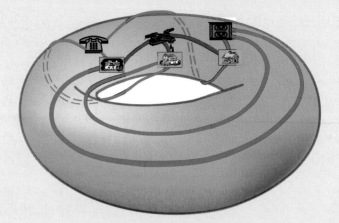

▶ CHINESE POSTMAN PROBLEM
(page 67)

One of the possible Eulerized graphs is shown here. As you can see, there were four vertices with odd numbers of edges (in red), which we have connected together with the green paths as shown. It is now possible to trace your way around the entire diagram and finish back where you started, thus creating an Eulerian circuit.

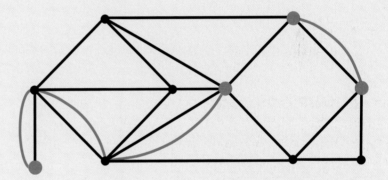

▶ DOING THE ROUNDS
(page 68)

The graph corresponding to this problem is shown. Because all its vertices are even there must be an Eulerian circuit, and it won't be difficult to find one.

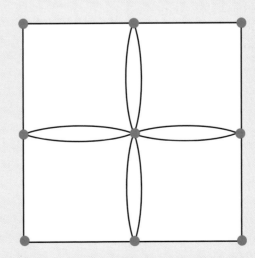

TRAVELING SALESMAN PROBLEM (page 69)

The best possible solution is shown at the top right, with the salesman travelling a total distance of 29 units.

In general, how many different routes would you have to investigate if the salesman had to visit "n" towns, including his home? The raw answer is given by:

$(n-1)! = (n-1)$ factorial $=$ all the numbers from 1 to $(n-1)$ multiplied together.

Given that half of these routes are merely reversals of the other half, the actual total is half this. As "n" increases, the value of half $(n-1)!$ blows up massively, as the following table demonstrates:

Growth of $(n-1)!/2$	
n	$(n-1)!/2$
3	1
4	3
5	12
8	2,520
12	19,958,400

Although there are only 12 routes to investigate in this particular puzzle, you can appreciate how it becomes very tricky for, say, 200 towns, even if you have a staggeringly powerful computer. Unfortunately, there is no known efficient algorithm that will solve this puzzle. However, it is possible to use commonsense rules that can give us a near-enough answer.

For example, one rule of thumb is called the Nearest Neighbor Algorithm. This rule states that whenever you arrive at a town, you always choose to travel to the (currently unvisted) town that is the closest. Using this rule in this puzzle, we obtain a route totaling 30 distance units (see illustration top left). This isn't as good as the perfect solution of 29, but it is pretty close considering that the worst result possible is 54.

CHOP CHOP (page 70)

No. The division of cake 1 and cake 3 was equal, but the red pieces in cake 2 were bigger. If the number of chords (cuts) is even and equal to 4 or more, the areas (pieces) are always equal. Otherwise, there is a slim chance (one chord must go through the center of the circle), but most likely not.

This puzzle was inspired by the "Pizza Problem" discovered by L.J. Upton in 1968; proven by Larry Carter and Stan Wagon in 1994.

(L.J. Upton, Problem 660, *Mathematical Magazine*, 1976; Stan Wagon, Problem 83, *The Mathematical Intelligencer*, 1983)

▼ CATERING FOR THE CHILDREN (page 71)

Puzzle 1 The maximum number of pieces into which the three cakes can be cut by 3, 4, and 5 cuts (a total of 12 straight line cuts) is 7, 11, and 16 pieces respectively, a total of 34 pieces. So the answer is yes.

This solution can be considered the minimal "best" solution, but there can be other solutions, if more than two cuts (lines) are allowed to meet at a point.

For example, cutting the cakes by 2, 4, and 6 cuts into 4 (max), 8, and 22 (max) pieces respectively.

This problem is a simple example from a branch of mathematics called combinatorial geometry, in which there is a fascinating interplay between shapes and numbers.

Puzzle 2 With the requirement of cutting the cakes into identical pieces we have to cut each cake radially from the center into 12 pieces, making 36 pieces altogether (in which case there will be pieces of cake for you and me as well).

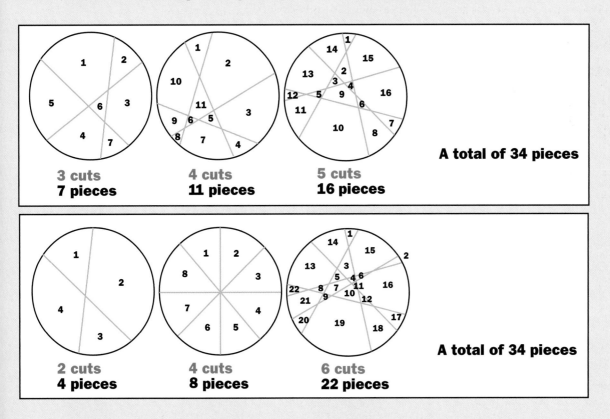

3 cuts
7 pieces

4 cuts
11 pieces

5 cuts
16 pieces

A total of 34 pieces

2 cuts
4 pieces

4 cuts
8 pieces

6 cuts
22 pieces

A total of 34 pieces

▼ COORDINATE CRAFT (page 74)

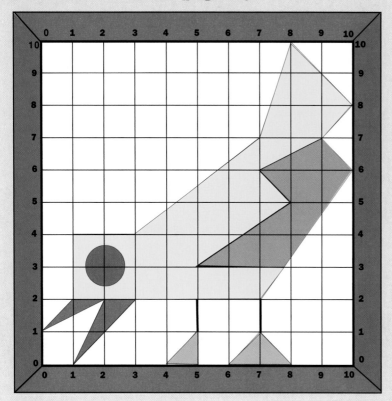

▼ ROYAL LINEAGE (page 75)

There are 13 distinct routes as shown.

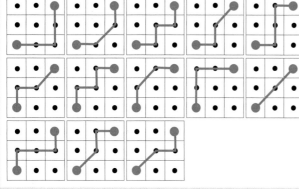

▼ REGAL PROCESSION (page 75)

There are nine distinct routes
as shown.

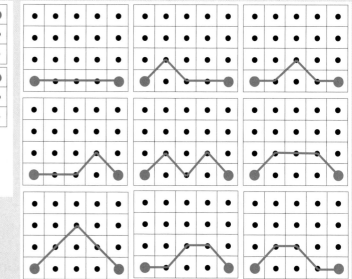

THINK OF A NUMBER (page 76)

Did you think about a kangaroo eating an orange in Denmark?

Don't be surprised. You are among the 98% of people who will answer the same. If you are among the other 2%, your answer might be an iguana eating an apple in Djibouti. Can you find any other solutions?

▶ MAKE A MEAL OF THIS (page 77)

The diagonal lines in the triangle represent the answers for 1 to 11 cuts for each of the three cases.

The structure of the triangle is the following: Each number in the triangular array is obtained by adding the number directly above and to the left together with the number that's directly above it two rows up (see examples shown).

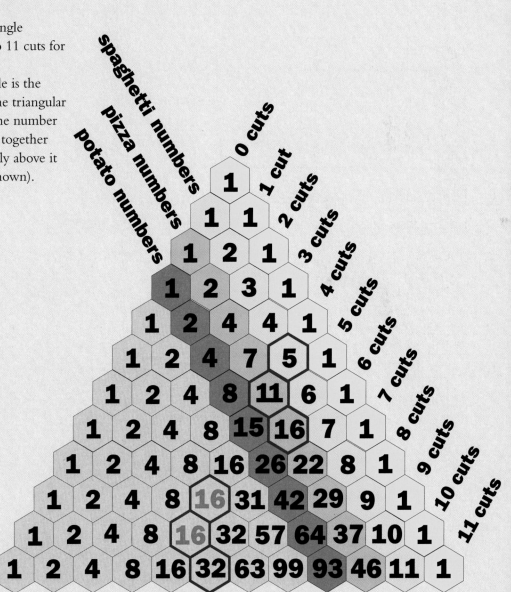

▶ CIRCULAR CONVERGENCE (page 78)

We can conclude that the area of the quarter circles is 9π units and the area of the semicircles is 8π units, and therefore the ratio of quarter circles to semicircles is 9:8. But subtracting the area of the red overlaps from both quarter circles and semicircles changes the ratio to an unknown quantity. It is therefore not possible to determine the ratio between the yellow and blue areas.

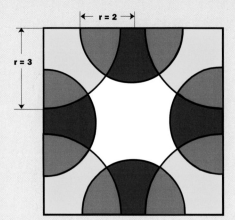

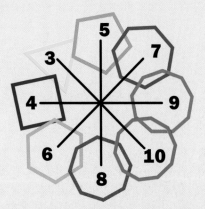

◀ POLYGON BRACELET (page 79)

Opposite polygons add up to 13.

▶ TRIANGLE TEASER (page 80)

You have to establish a systematic procedure to find all the triangles and to be sure not to miss one:

Single triangles: 1, 2, 3, 4, 5, 6, 7, 8, 9 (9)

Two joined triangles: 2–3, 2–6, 3–7, 5–9, 6–7, (5)

Three joined triangles: 2–3–4, 3–4–7, 3–7–8, 4–5–9, 5–9–8, 6–7–8 (6)

Four joined triangles: 1–3–4–5, 1–3–4–7, 1–4–5–9, 2–3–4–5, 6–7–8–9 (5)

Five joined triangles: 2–3–4–6–7, 2–3–6–7–8 (2)

Six joined triangles: 0

Seven joined triangles: 1–3–4–5–7–8–9 (1)

Eight joined triangles: 2–3–4–5–6–7–8–9 (1)

Altogether there are 29 triangles.

FLUID DYNAMICS (page 81)

Water will start pouring out of the top hermetically sealed compartment through the right tube, making the pressure in it lower. This causes a fountain of colored water to come up through the left thin tube, raising the water level in the upper container.

▼ GANYMEDE CIRCLE 1 (page 82)

For clarity, distances are always measured clockwise around the circle.

Distance	Measure clockwise
1	A to B
2	B to D
3	A to D
4	D to A
5	D to B
6	B to A
7	Entire circle

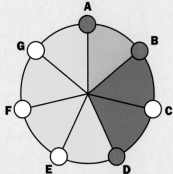

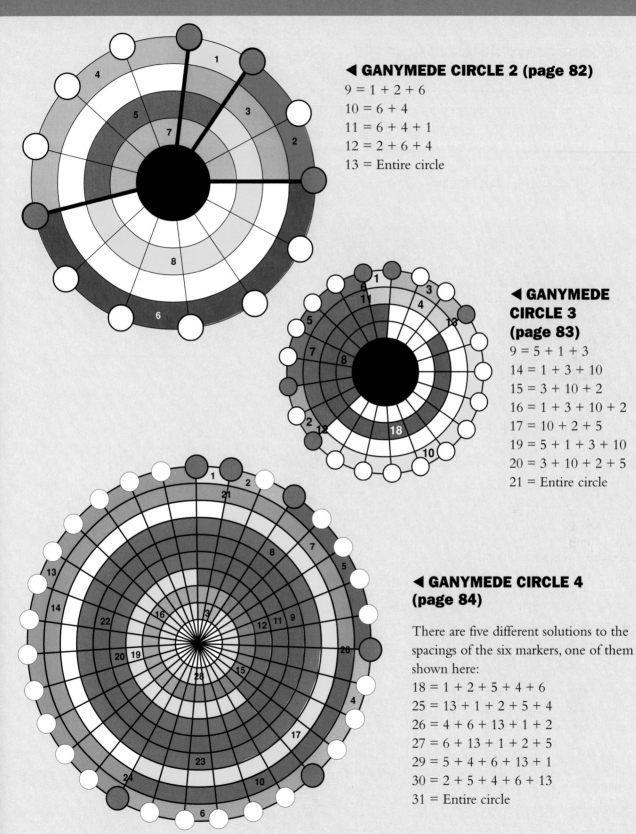

◀ **GANYMEDE CIRCLE 2 (page 82)**

$9 = 1 + 2 + 6$
$10 = 6 + 4$
$11 = 6 + 4 + 1$
$12 = 2 + 6 + 4$
$13 =$ Entire circle

◀ **GANYMEDE CIRCLE 3 (page 83)**

$9 = 5 + 1 + 3$
$14 = 1 + 3 + 10$
$15 = 3 + 10 + 2$
$16 = 1 + 3 + 10 + 2$
$17 = 10 + 2 + 5$
$19 = 5 + 1 + 3 + 10$
$20 = 3 + 10 + 2 + 5$
$21 =$ Entire circle

◀ **GANYMEDE CIRCLE 4 (page 84)**

There are five different solutions to the spacings of the six markers, one of them shown here:

$18 = 1 + 2 + 5 + 4 + 6$
$25 = 13 + 1 + 2 + 5 + 4$
$26 = 4 + 6 + 13 + 1 + 2$
$27 = 6 + 13 + 1 + 2 + 5$
$29 = 5 + 4 + 6 + 13 + 1$
$30 = 2 + 5 + 4 + 6 + 13$
$31 =$ Entire circle

▶ COLORFUL COMBINATION LOCK
(page 85)

The answer is the red sections.

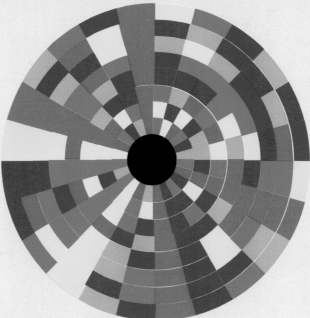

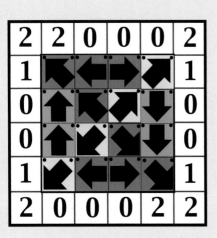

◀ MAGIC ARROWS: GAME 1
(pages 86–87)

▼ MAGIC ARROWS: GAME 2
(page 88)

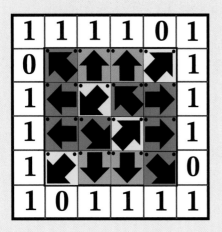

▼ MAGIC ARROWS: GAME 3
(page 89)

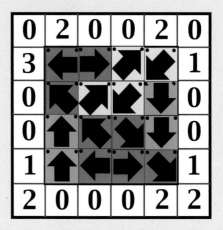

▶ IN YOUR AREA (page 90)

The answer is 32 regions.

We can prove this using Euler's formula, which we met on page 65. When two circles overlap, they cross at two places at most. This creates $n(n - 1)$ intersections (or vertices) for "n" circles.

Similarly, each edge of the "n" circles is cut up into $2(n - 1)$ edges, which equals $2n(n - 1)$ edges in total.

Euler's formula states that:

Faces = Edges − Vertices + 2

So, Faces = $2n(n - 1) - n(n - 1) + 2$
$$= n(n - 1) + 2$$
$$= n^2 - n + 2$$

Since $n = 6$ here, Faces = $6^2 - 6 + 2 = 32$, as given above.

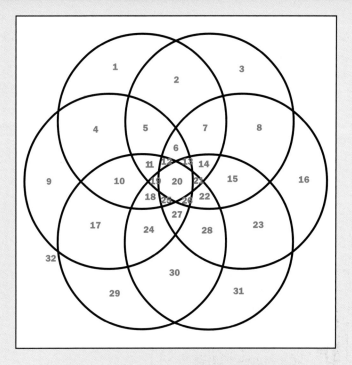

▶ CLOVER COVERAGE (page 91)

Area of the square BDEF
$$= BF^2$$
$$= (2 \times BA)^2$$
$$= (2 \times AC)^2 \text{ by symmetry}$$
$$= [2 \times (OA + OC)]^2$$

Since OA = 1 (radius of circle) and OC = $\sqrt{2}$ (Pythagoras)

Area BDEF = $[2 \times (1 + \sqrt{2})]^2 = 23.31$ units²

Area of clover leaf
$$= \tfrac{3}{4} \text{ of circle + square OGCH}$$
$$= \tfrac{3}{4}(\pi \times 1^2) + 1^2$$
$$= \tfrac{3\pi}{4} + 1 = 3.36 \text{ units}^2$$

Hence, 3 leaf clover = 10.08 units²
      4 leaf clover = 13.44 units²

Area of clover = $(16 \times 10.08) + (9 \times 13.44)$
$$= 161.28 + 120.96$$
$$= 282.24 \text{ units}^2$$

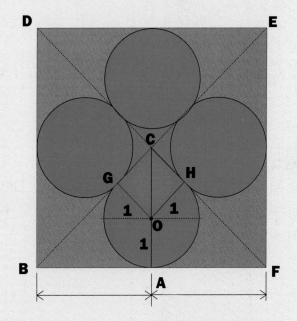

Area of entire field = 25 × 23.31 (from above)
      = 582.75 units²

Hence, proportion that is clover = 282.24/582.75 = 48.43%

▼ THE EIGHTEENTH AMENDMENT (pages 92–93)

No matter how the points are distributed, it will be impossible to place the 18th point (proven by Berlecamp, Graham, and Warmus in 1970). There are 768 different 17-point solutions (counting reversals as equivalent), one of which is shown.

This beautiful and unusual problem first appeared in *One Hundred Problems in Elementary Mathematics* by the Polish mathematician Hugo Steinhaus, and was later extensively covered by Gardner, Conway, Baxter, and others.

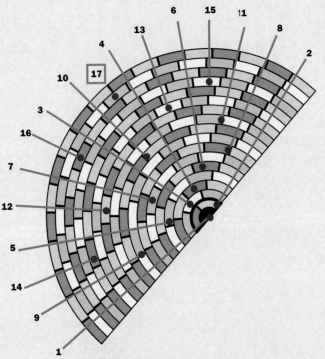

▼ ACCESS ALL AREAS (page 94)

The area of the hexagon is 50% larger than the area of the equilateral triangle, as shown here.

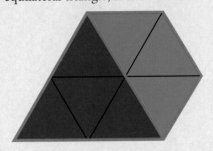

TO THE MAX (page 95)

All the triangles have the same area, since they have equal bases and the same perpendicular height.

ONE FOR THE BIRDS (page 96)

Three red birds (R) and two blue birds (B), as demonstrated below.

$R - 1 = B$

$3 (B - 1) = R$

Hence, $3B - 3 = B + 1$

$2B = 4$

$B = 2$

$R = 2 + 1 = 3$

THE SAFE SAFE (page 97)

There are 26 letters in the alphabet and four letters will be chosen for the combination lock, so you multiply $26 \times 25 \times 24 \times 23 = 358,800$ in order to obtain the number of permutations.

So, if it takes 5 seconds to try each permutation, it would take nearly 500 hours to find the right one!

▼ WINDOW OF OPPORTUNITY (page 97)

The new window is shown here.